EXTRACTIVE RESERVES IN BRAZILIAN AMAZONIA

To my mother

Extractive Reserves in Brazilian Amazonia

Local resource management and the global political economy

CATARINA A.S. CARDOSO
University of Westminster

LONDON AND NEW YORK

First published 2002 by Ashgate Publishing

Reissued 2018 by Routledge
2 Park Square, Milton Park, Abingdon, Oxon OX14 4RN
711 Third Avenue, New York, NY 10017, USA

Routledge is an imprint of the Taylor & Francis Group, an informa business

A Library of Congress record exists under LC control number: 2001098619

ISBN 13: 978-1-138-73786-0 (hbk)
ISBN 13: 978-1-138-73782-2 (pbk)
ISBN 13: 978-1-315-18515-6 (ebk)

Contents

List of Figures and Maps

List of Tables

Acknowledgements

During the researching and writing of this book many people have given me support and advice. I would like to start by thanking Dr. Anthony Hall, for his most valuable comments throughout the various stages of my research, and for providing me with crucial contacts in Brazil. I also warmly thank Professor Judith Rees for her assessment and comments on my work. Dr. Mark Pennington's frank and engaged comments and his strong encouragement at the most difficult points of this research were of immense help – I am most grateful to him.

In Brazil, I am indebted to many people whose help was invaluable for my field research. Many offered me accommodation, provided me with free access to computer facilities, key contacts for my research and received me like a close friend. In particular, I would like to express my gratitude to Tony Waugh, Eleanora Dobin and her family, Ana Elisa and Zé, Ana's sisters Glaucia, Regina and Joynes, and their family, Manoel and Nelia, Dna Elais, Julia Feitosa, Carlos Carvalhas, Dau, Pingo and his family, in particular Haylen, Branca, Lurdes, Haila and Charly, and José from *colocação* Cumarú on *seringal* Icuriã. I would also like to thank Dr. Reginaldo Fernando Castela from the Federal University of Acre and Dr. Donald Sawyer and Bruno Panoccheschi of the ISPN (Instituto Sociedade, População e Natureza). Apart from logistic support they also supplied important contacts. Special thanks go to all the policy-makers and rubber tappers I interviewed and, in particular, to those extractivists who received me in their houses.

In the UK, I am particularly grateful to Iris Hauswirth, for her comments on the theory chapter and for her patient technical support, and to Eleanor O'Gorman, Polly Mohs and Hans Stolum, for helping me set the foundations of this research.

This work would not have been possible without the financial support of JNICT, Junta Nacional de Investigação Científica e Tecnológica, Calouste Gulbenkian Foundation, London School of Economics and the British Federation of Women Graduates.

Finally, I owe special thanks to Holger.

List of Abbreviations

AMOREAB	Associação de Moradores da Reserva Extrativista Chico Mendes Região de Assis Brasil (Association of Inhabitants of the Extractive Reserve Chico Mendes Region of Assis Brasil)
AMOREB	Associação de Moradores da Reserva Extrativista Chico Mendes Região de Brasiléia (Association of Inhabitants of the Extractive Reserve Chico Mendes Region of Brasiléia)
AMOREX	Associação de Moradores da Reserva Extrativista Chico Mendes Região de Xapuri (Association of Inhabitants of the Extractive Reserve Chico Mendes Region of Xapuri)
BIRD	Banco Internacional Para a Reconstrução e o DesenvolvimentoInternational Bank for Reconstruction and Development
CAEX	Cooperativa Agro-Extrativista de Xapurí (Agro-extractivist Co-operative of Xapurí)
CDEA	Commission on Development and Environment for Amazonia
CEC	Commission of the European Communities
CEDI	Centro Ecumênico de Documentação e Informação (Ecumenical Centre for Documentation and Information)
CNDDA	Campanha Nacional de Defesa e pelo Desenvolvimento da Amazonia (National Campaign for the Defence and Development of Amazonia)
CNPT	Centro Nacional para o Desenvolvimento Sustentado das Populações Tradicionais (National Centre for the Sustainable Development of Traditional Populations)
CNS	Conselho Nacional de Seringueiros (National Council of Rubber Tappers)

CNS-RVAP	Conselho Nacional dos Seringueiros – Regional do Vale Acre-Purús (National Council of Rubber Tappers – Regional River Valley Acre and Purús)
CONTAG	Confederação Nactional dos Trabalhadores na Agricultura (National Federation of Agricultural Workers)
CPR	Common Pool Resource
CTA	Centro dos Trabalhadores da Amazônia (Amazon Workers' Centre)
CUE	Comissão da União Europeia Commission of the European Union
CUT	Central Única do Trabalho (Unified Labour Centre)
CVRD	Companhia Vale do Rio Doce Vale do Rio Doce Company
ELI	Environmental Law Institute
ER	Extractive Reserve
ERCM	Extractive Reserve Chico Mendes
FAO	United Nations Food and Agricultural Organisation
FNMA	Fundo Nacional do Meio Ambiente (National Environmental Fund)
FoE	Friends of the Earth
G7	Group of 7
GoB	Government of Brazil
GTA	Grupo de Trabalho da Amazônia (Amazon Working Group)
IBAMA	Instituto Brasileiro do Meio Ambiente e dos Recursos Renováveis (Brazilian Institute of the Environment and Renewable Resources)
IDB	Inter-American Development Bank
IEA	Instituto de Estudos Amazônicos e Ambientais (Institute of Amazon Studies)
INCRA	Instituto Nacional de Colonização e Reforma Agrária (National Institute for Colonisation and Agrarian Reform)
INPE	Instituto Nacional de Pesquisas Espaciais (National Institute for Space Research)

ITTA	International Tropical Timber Agreement
ITTO	International Tropical Timber Organisation
LPNMA	Lei de Política Nacional do Meio Ambiente Law of the National Policy for the Environment
MDB	Multilateral Development Banks
NGO	Non-Governmental Organisation
PAE	Projeto de Assentamento Extrativista (Extractivist Settlement Project)
PMACI	Programa de Proteção do Meio Ambiente e às Comunidades Indígenas (Programme for the Protection of the Environment and Indigenous Communities)
PNMA	Plano Nacional do Meio Ambiente (National Environment Program)
POLAMAZONIA	Programa de Polos Agropecuários e Agrominerais da Amazônia (Programme of Agro-Livestock and Agro-mineral Poles for Amazonia)
POLONOROESTE	Programa de Desenvolvimento Integrado do Noroeste do Brasil (Integrated Development Programme for the North-West Region of Brazil)
PP-G7	Pilot Programme for the Conservation of Brazilian Rainforests
PRODEX	Programa de Apoio ao Desenvolvimento do Extrativismo (Assistance Programme for the Development of Extractivism)
PT	Partido dos Trabalhadores (Workers' Party)
RESEX	Pilot Programme sub-project on Extractive Reserves
SAE	Secretaria de Assuntos Estratégicos (Secretariat for Strategic Affairs)
SEMAM	Secretaria Especial do Meio Ambiente (Special Secretariat for the Environment)
SUDAM	Superintendência do Desenvolvimento da Amazônia (Superintendency for the Development of Amazonia)
TFAP	Tropical Forest Action Plan

TORMB	Taxa de Organização e Regulamentação do Mercado da Borracha (Tax for the Organisation and Regulation of the Rubber Market)
UDR	União Democrática Ruralista (Farmers' Democratic Movement)
UNCED	United Nations Conference on Environment and Development
UNDP	United Nations Development Programme
UNEP	United Nations Environmental Programme
WB	World Bank
WCED	World Commission on Environment and Development
WRI	World Resources Institute
WRM	World Rainforest Movement
WWF	World Wide Fund for Nature

Glossary of Portuguese Terms

Aforamento:	"A lease from the state for extractive rights. The boundaries of *aforamentos* were notoriously imprecise, but these usufruct rights were later transformed into de facto land rights as large owners began to sell aforamentos areas for ranching" (Hecht and Cockburn, 1989, p.271)
Aviamento:	Debt bondage system characteristic of the rubber trade in Amazonia
Benfeitorias:	Improvements made to the land
Capoeira:	Secondary vegetation
Colocação:	Rubber stand
Colônia:	Agricultural Settlements
da margem:	Near the riverbank
Empate:	Stand-off
Estradas de seringa:	Rubber trails
Fazenda:	Ranch, large holding or plantation
Lideranças:	Leadership, former tappers usually living outside the reserve many led the resistance process against the ranchers
Município:	Municipality
Patrão:	The patron, the person to whom one sells and from whom one might be able to ask for favours every once in a while (Hecht and Cockburn, 1989, p.272)
Portaria:	Internal decree
Posseiro:	Untitled occupier of a plot
Seringais:	Rubber estates
Seringal:	Rubber estate
Seringalista:	Rubber estate owner, rubber baron
Terras devolutas:	"Unsurveyed lands of the state. These can be lands that were never leased or can be land whose leases or claims have not been renewed" (Hecht and Cockburn, 1989, p.272)

Introduction

Extractive reserves are one of Brazil's main policy initiatives for promoting sustainable development in the Amazon rainforest. The result of local communities' struggles for control over their natural resources, and worldwide concern with the conservation of Amazonia, extractive reserves also represent an arena for exploring linkages between local resource management and the global political economy. References to the importance of such linkages are frequent, but they remain poorly understood. This book addresses this lacuna through an analysis of the formation, and institutional sustainability of the extractive reserves.

In the last few years, it has become common knowledge that community participation in the management of natural resources is essential to address environmental degradation at the local level, and ensure the livelihoods of poor people. This is a relatively recent development because in the recent past, traditional communities were considered unable to conserve their natural resources, either because they did not possess the necessary 'scientific' knowledge or because they were supposedly trapped in a 'tragedy of the commons' situation. By contrast, local resource management is now emphasized in both the environmental and the development literature. The concept of sustainable development offered by the World Commission on Environment and Development, which was pivotal in giving political leverage to the notion that development goals need to incorporate environmental concerns, illustrates this new prominence of community control over their resources:

> The pursuit of sustainable development requires a political system that secures effective participation in decision making ... This is best secured by decentralising the management of resources upon which local communities depend, and giving these communities an effective say over the use of these resources ... (UN World Commission on Environment and Development, 1987, p.63,65).

The idea that many environmental problems are global and that therefore they need to be addressed at the international level has also gained considerable recognition in the last 10 years. Growing scientific knowledge, a series of important environmental disasters, and the action of

NGOs in publicising the worldwide impacts of pollution of the atmosphere, deforestation, and species destruction contributed to make these problems well known to the general public and to place them firmly on the international politics agenda. This process culminated in the United Nations Conference on Environment and Development, which took place in 1992, in Rio de Janeiro. Here, developed and developing countries' governments set out to discuss global environmental issues, such as climate change, biodiversity depletion, and deforestation. The latter was, however, the only issue on which no binding agreement could be reached.

Studies about the livelihoods of local communities in the Third World, frequently point out that these communities' options are largely influenced by the global political economy, of which global environmental issues are now part. The world economic structure and international environmental policies help to shape national policies, which in turn affect local populations. Likewise, analyses of global environmental problems tend to note that they can only be tackled if development issues at the local level are also addressed. Nowhere is this clearer than in the case of deforestation. In Brazilian Amazonia, forest clearing has been partly caused by landless peasants who moved to the Amazon region in search of a better life, totally unaware of the agricultural limitations of the soil when cleared of its forest cover. Similar situations exist elsewhere, in the forests of the Congo Basin and of South East Asia.

Notwithstanding the interest in such concerns, the linkages between local and global environmental issues and their corresponding socio-economic processes remain poorly understood. Most studies tend to focus on either global or local questions rarely exploring the dynamic process of interaction between local and global processes. Analyses of global issues largely ignore local processes and research on local resource management discusses only global developments that are *detrimental* for local communities. In exploring local resource use, the present study gives equal attention to local, national and global political processes that influence communities' use of the forest.

This book examines the interrelationship between local institutional factors and the wider socio-political and economic context with regard to forest management. It centres on the management of the commons and explores two central questions: first, what influences the creation of common property regimes and, second, what affects their capacity to ensure sustainable development. The focus is on extractive reserves, community institutions developed in the 1980s by Amazonian forest dwellers, rubber tappers, amid considerable international concern with deforestation in the region.

Analyses of the commons are particularly important because many people in the developing world depend to a higher or lesser extent on shared resources. This is likely to continue for the foreseeable future (Dasgupta, 1996). It is now well established that, contrary to the 'tragedy of the commons' scenario, users of common resources do not inevitably destroy them. The commons can, and many have been, sustainably used for centuries. But not all commons are conserved and what distinguishes those that are from those that are not remains a central question for those concerned with either development or environmental conservation.

The theory of institutional choice has come to be the most coherent and influential theory for explaining why and how some groups of individuals ensure the conservation of their common resources. While yielding important insights into the design of local institutions capable of harmonising the use of common resources, institutional theory, however, has mainly focused on local determinants, such as the characteristics of the group of users. Within this theoretical framework, analyses of the wider socio-political and economic context where local resource users are embedded, and the influence they have in the design of such institutions are very recent. By contrast, the literature on 'environmental action' and 'participation', which also addresses many issues pertaining to local resource management, has paid considerable attention to the role of outsiders and the wider socio-economic context. Much of this research has, however, concentrated on the struggle between local resource users and outsiders who attempt taking over their resources, often with 'development' objectives in mind or to preserve natural resources from a potential 'tragedy'. Which local arrangements ensure the conservation of the common resources once the struggles to gain control over them are over, is a question that frequently remains unanswered. The way in which the external context continually interacts with resource users and shapes their institutions, providing positive as well as negative influences has in general been a neglected issue.

The present analysis of extractive reserves seeks to combine these two strands of research to understand how local resource management can be influenced by the external context. By and large, the emphasis of many studies in either institutional theory or 'political ecology' is on showing that local communities can manage their natural resources and that they should be given sufficient power to do so. This book moves beyond this: it examines the relationship between outsiders and local users, both in the context of struggles for gaining control of resources and once resource users gain the right to manage their commons. To a higher or lesser extent, most common property regimes are embedded in the external context and thus influenced by a variety of external factors. The pastureland tenure regimes of Eastern Africa, for example, have been often affected by national wildlife conservation policies (Lawry, 1990; Lane 1998); the Mexican *ejidos* by the macro-economic policies of the Mexican government (Yetman, 1998), and the Chipko people, who also hold their forests in common, have formed alliances with international environmental organisations, and are thus nested in the international as well as in the Indian national setting (Rangan, 1993). The framework developed here to explore extractive reserves can be applicable to the analysis of common property institutions anywhere in the world. Therefore, while the focus of the research is quite specific – as summarised below – it is hoped that its analysis will have wider relevance.

This study focuses on the history and present characteristics of extractive reserves, which are areas, set aside for both the conservation of the natural resource base and its utilisation by local populations. The first reserves were set up in 1990, and there are now 16 federal extractive reserves (12 in Amazonia and 4 marine reserves), and 21 state administered reserves, covering a total area of nearly 45,000 sq. km, and benefiting a population of approximately 50,000 people. The creation of 21 new reserves is envisaged over the next few years. The extractive reserves' concept is innovative from at least three perspectives. First, it is virtually the first item in the Brazilian legislation that recognises common property rights to natural resources. Second, it support the rights of Amazonian local population to their lands and also to their mode of production – until the creation of the reserves government policy had concentrated mainly on other 'modern' and often environmentally destructive activities, largely ignoring and sidelining local populations. Third, the extractive reserve concept combines conservation with productive use of natural resources, thus moving beyond the view that to conserve natural areas local populations need to be removed. These three features alone make the significance of extractive reserves go beyond Brazil, since the recognition of common property, support for traditional activities, and combination of use and conservation are recurrent themes in debates about forests worldwide. The application of the concept of extractive reserves has been discussed in various other countries.

The significance of extractive reserves, however, lies also in the fact that they encapsulate many interdependent local and global issues. Tropical deforestation has received considerable attention in the international arena, in particular that occurring in Brazilian Amazon, as the largest remaining tropical rainforest. Its destruction can intensify the 'greenhouse effect' and considerably reduce the Earth's biodiversity reserves, two global problems in their own right. Extractive reserves, in turn, are forest areas used and largely managed by forest dwellers, who obtained rights to their land after fighting for them over nearly two decades. Theirs was a fight for secure land rights, not global issues. However, it is widely acknowledged that international interest in deforestation in Amazonia was a determinant factor in the creation of the reserves. Presently, the reserves continue to receive support from non-local actors, including international NGOs, such as the World Wide Fund for Nature (WWF) and by an initiative of the G7 countries, the Pilot Programme for the Protection of Brazilian Rainforests. This support largely stems from the belief that these institutions contribute to the resolution of the globally important problems mentioned above. Whether they fulfil this goal is one of the questions this book attempts to answer.

Given the amplitude of the topic and the complexity of the issues involved, undertaking the study of several communities' relationships with the wider context would have necessarily meant cutting down on detail and depth of analysis. Consequently, the present research focuses on one particular reserve, the Extractive Reserve Chico Mendes, which is the largest reserve in Brazil, both in terms of area and population, and where the movement that led to the creation of extractive reserves originated. This case, it is hoped, will shed light on many issues regarding local resource management by communities who live neither in isolation nor in permanent struggle against outsiders.

For understanding both the local factors that defined the rubber tappers' institutions and their wider context, I conducted in depth interviews with both rubber tappers and policy-makers who have participated in Amazonian policies for the last 20 years. For five months I visited the rubber tappers' houses in the forest, interviewing them at length about their relationship with other tappers in the area, with government officials and development workers, and above all with the forest and its resources. Before and after visiting the rubber tappers, I interviewed members of the government, the parliament, NGOs and academics about Amazonia. The aim was to understand whether and how the events that took place at the time of the creation of the reserves had influenced them, and how was their situation in the wider context in the 1990s.

The structure of the book

Chapter 1 develops an analytical framework to examine the formation and robustness of common property regimes. This framework is composed of three parts. The first reviews a set of criteria for assessing the robustness of a common property regime, that is, the regime's capacity to secure the conservation of the resource during long periods. The second part examines potentially significant factors in the development of a robust regime, focusing on those pertaining to the characteristics of the resource and the resource users. The third part brings into the analysis factors arising from the external context, such as government policies and socio-political changes taking place in society at large, which influence both the capacity of regimes to secure the conservation of resources and the development of robust regimes.

As the aim of this book is to examine the commons in relation to the external context, Chapter 2 reviews how the national and international context in which rubber tappers and other Amazonian commoners are embedded, has evolved since the 19th century down to the present day. The rubber tapper population was formed at the end of the 1800, when the rubber boom – largely triggered by industrialisation and scientific discoveries outside Amazonia – attracted landless peasants from the Northeast of Brazil to the Amazon rainforest. After a period of considerable isolation (1920-1965), the region was again the focus of attention in the 1970s and 1980s. National government policies for development and international concerns with deforestation during these decades not only formed the context in which the tappers' struggle for land rights took place, but, as I argue in subsequent chapters, also helped shape the current institutional characteristics of the extractive reserves.

Chapter 3 examines the evolution of the rubber tappers' property rights regimes from the time of their arrival in Amazonia until the establishment of extractive reserves. Initially, the rubber tappers worked in conditions of semi-slavery on large and privately owned rubber estates. From the 1920s onwards working conditions considerably improved and in the 1960s many tappers became 'independent', managing their resources in loose common property regimes. The evolution of the rubber tappers' institutions from their 'independence' until the establishment of extractive reserves in the late 1980s is linked to a range of national and international developments. After arguing that not all these developments were detrimental to the rubber tappers, I show how a complex range of historical, economic and political factors shaped the tappers' reaction to either friendly or unfriendly external developments.

Chapter 4 reviews the legislative and formal aspects of extractive reserves, and presents a brief overview of the features of the Pilot Programme sub-project on extractive reserves, the principal international initiative providing support for reserves. In the discussion about the relationship between local resource users and outsiders, the comparison between the formal and informal aspects of the reserve is particularly useful. It shows how difficult it is in practice to achieve the dual aims of providing support for commoners without impose on them alien institutions. The examination of the Pilot Programme, in this and the last chapter of the book, provides additional evidence of such difficulties.

Chapter 5 examines the informal institutional arrangements of an extractive reserve, comparing them with the formal set up described before[1]. The chapter attempts to unpack the reasons behind the reserve inhabitants' resource management system and assesses whether such systems can ensure the long-term conservation of the forest given the current national and international setting. This discussion is expanded in Chapter 6, which brings together the different themes of this study, to cover the broader issue of sustainable development – instead of only conservation – and implications relating to the role of the global political economy in the lives of rural commoners. The policy implications of the analysis are dealt with in the concluding chapter.

Note

[1] As Figure 5.1 shows, the Chico Mendes Reserve (ERCM) is located in the state of Acre, in the southeast of Amazonia. It was one of the first extractive reserves to be established and covers an area of nearly 10,000 sq. km where approximately 7,000

extractivists live (Ibama, c2001). Although the process that led to the establishment of extractive reserves began in an area that is now part of the ERCM, many tappers living now in the area of the ERCM did not participate in the fight for landed property rights. The study of this reserve thus provides a larger variety of situations than it would have been possible in smaller and more homogenous reserves. At the same time, it provides a clear insight of how the situation of the other reserves may be and which factors need to be considered in policy making in this respect.

1 The Conceptual Framework: 'Common Property Institutions'

Introduction

A central component of local resource management is finding sustainable ways of sharing natural resources. Most rural communities in the developing world depend to a higher or lesser extent on common resources: the entire livelihood of many pastoral communities, for example, is contingent on jointly used pastures (Lane, 1998); other groups obtain their income from private plots of land or from salaried work, but nevertheless share essential resources such as water and firewood (Dasgupta, 1996). Occasionally, the destruction of common resources might be beneficial for its users, but in the large majority of cases, rural communities suffer dramatic losses when this happens. Yet, the conservation of common resources can be very problematic.

A useful concept for identifying the difficulties of sustainably sharing resources is that of 'common-pool resource'. The defining feature of common-pool resources (CPRs) is that several individuals can use them in common, but only on condition that their joint use is below a certain threshold of use: if that threshold is crossed the resource risks depletion. Broadly speaking, there are two lines of thought regarding the conservation of CPRs. One claims that resource users are unable to ensure the sustainable use of the resource by themselves and, therefore, to prevent the depletion of a CPR, there are only two solutions: state control or privatisation of the resource. The other argues that resource users *can* ensure the sustainable use of CPRs through the development of common property institutions, but that whether they do so depends on a wide range of factors. After outlining these two arguments, the present chapter develops a framework for examining what influences the sustainable use of resources held as common property, which includes both internal factors pertaining the characteristics of the resource and its users, and external factors nested in the national and international context.

Common-pool resources

A common-pool resource is a 'a natural or man-made resource system that is sufficiently large as to make it costly (but not impossible) to exclude potential beneficiaries from obtaining benefits from its use' (Ostrom, 1990, p.30). Common-pool resources are often confounded with 'public goods', but although there are important similarities between them, there are also crucial differences, which is why mechanisms for providing public goods are frequently inappropriate to ensure the conservation of CPRs.

The use of CPRs, like that of public goods, is non-rival, thus several individuals can jointly use a common-pool resource without this imposing any costs on fellow users. For instance, several persons can share a forest, as the rubber tappers and other Amazonian forest dwellers do, without interfering with each other. However, public goods have a higher level of 'jointness' than CPRs, for the benefits of the latter are non-rival only up to a certain level of use, and only if used under certain conditions (Oakerson, 1986; Ostrom and Ostrom, 1978). In the case of a fishery, for example, the resource may be jointly used provided that only fish above a certain size is caught; in the case of a forest, a 'rest' period may be necessary to replenish the resource or users may have to restrict the variety of products they extract from the forest.

Both public goods and CPRs share the feature 'difficult to exclude from benefits', but whereas in the case of pure public goods difficulties are of such a magnitude as to make exclusion impossible, in the case of CPRs exclusion is difficult but feasible. One of the main reasons why exclusion from CPRs' benefits is difficult is that CPRs are often indivisible, but contrary to what is the case with public goods, most CPRs can be privatised, in which case exclusion of non-owners is made easier. Moreover, many of the benefits of a CPR are private goods and can thus be handled as such. One characteristic of CPRs is that the resource system (e.g. the forest, the fishery, the pasture) can be used in common, and, to a certain extent, it is like a public good, whereas the resource units (e.g. the fruits from the forest, the fish, the grass eaten by each animal in a pasture) are pure private goods. Consumption of the resource units is rival and exclusion is easy.

The conservation of a shared CPR presents problems similar to those involved in the provision of a public good. The conservation of a CPR depends on the harmonisation of the actions of the different users of the resource; if all users act independently, that is, disregarding the impact of their actions on fellow users, the total use of the resource may be above its regenerative capacity. When deciding how many units to take from the resource system, users must bear in mind that although their individual use might be relatively small, when adding up the units all the resource users take, total extraction can be above the threshold level of use. To secure the sustainable use of the resource, users should therefore agree to restrict their own use of the CPR.

Yet, just as people have an incentive not to contribute to the provision of a public good (for once the good is provided they cannot be excluded from its benefits), users of a CPR have an incentive not to contribute to the conservation of the resource (for it is very difficult for the others to prevent them using the resource). Any user can free ride on the conservation efforts of the other users. Individuals may therefore fear that the other users will free ride on their conservation efforts, or think that they can free ride on the conservation efforts of the others. To secure the conservation of the resource it is thus necessary to find a means of securing that no one free rides on the others' efforts; otherwise, the resource can be depleted leaving everybody worse off (Runge, 1986; Ostrom, 1990).

There are three standard solutions to the free rider problem in relation to CPRs[1]. One is that the state takes control of the resource; the main argument behind this solution is that the conservation of a CPR is like a public good, and as such, the state only can provide it. The state, however, has often proved not to be up to the task (Bromley and Cernea, 1989). There are many possible explanations for this. The state often has other priorities than the conservation of natural resources; besides, its objectives are sometimes incompatible with the conservation of CPRs (Turner, Pearce and Bateman, 1994). Furthermore, even if genuinely committed to conservation, the state often lacks the capacity to monitor the use of the resource, and the knowledge of how best to manage it (Ostrom, 1990).

Another potential solution is to privatise the common-pool resource. As by definition CPRs tend to be indivisible, this solution implies that one entity – an individual or a firm – becomes the owner of the entire resource. Advocates of privatisation, market environmentalists, argue that if the resource is privately owned, the owner will be the sole beneficiary of conserving the resource and, therefore, will only deplete the CPR if the costs of doing so are lower than the benefits (Demsetz, 1967; Kwong, 1992). Privatisation of common-pools therefore prevents users from depleting them against their best interests, yet it fails to guarantee the conservation of natural resources. Moreover, privatising a CPR implies that at least some if not all of the previous users are excluded from its use, which can have a high social cost, especially if the resource users are poor as in developing countries they often are. This was what occurred with the rubber tappers: in the late 1970s, their jointly used forests were privatised (albeit not for conservation purposes), and unable to find alternative livelihoods most of them became considerably worse off.

Although privatisation of CPRs has important shortcomings and is often unfeasible, when looking for solutions for common pool resources some of the arguments of market environmentalists need to be considered. One such argument is that well-defined property rights are crucial for preventing the destruction of natural resources. Defining property rights reduces uncertainty – owners know what they can and cannot do with the resource, and they know who will benefit from their actions. Security of continuous access to the resource is an incentive for pursuing economically efficient actions because if economic agents are certain that the resource will be there in the future, they do not need to take full advantage of the CPR today at the expense of its long-term sustainability. They can refrain from extracting units from the common-pool, in the certainty that they will be the ones to benefit from their actions. Private property, however, is not the only property rights institution that satisfies these requirements.

Advocates of either state ownership or privatisation assume that joint users of a common-pool cannot overcome the free-rider problem and therefore that any CPR that is jointly used is in risk of depletion, an argument that was epitomised by Hardin (1968) in his seminal article 'The Tragedy of the Commons'. The tragedy Hardin refers to is the supposedly unavoidable depletion of a common pasture in spite of the fact that this is not in the interest of the herdsmen. According to Hardin, the herdsmen will not be able to stick to any agreement involving a limitation of their use of the pasture. Although all the herdsmen will be worse off by using the resource too much, 'each man is locked into a system that compels him to increase his herd without a limit – in a world that is limited' (Hardin, 1968, p.20). The system that supposedly compels each herdsman to overuse the resource is similar to the one faced by the prisoners in the prisoners' dilemma model. Each prisoner assumes that whatever decision the other prisoner takes, whether he confesses or hides the crime from the police, he is better off confessing; as both think in the same way, they both confess the crime and are worse off than if they had kept silent about it (Axelrod, 1984; Runge, 1986).

There is, however, substantial empirical evidence of commoners who have jointly used CPRs for a long time without depleting them, which suggests that users might be able to cooperate in spite of the free-rider problem. The free-rider problem is not as overpowering as often believed. The assumption that individuals free ride is based on the premise that individuals are primarily self-interested and thus when deciding on their level of use of the resource they will put their individual interest above the community's interest. The welfare of the community, however, may also be part of each user's individual welfare, in which case the incentive to free ride will not be dominant (Runge, 1986). Also, even if indifferent to the community's interest, users of a CPR can co-operate with each other in spite of their incentive to free-ride because they can communicate with each other and agree to jointly conserve the resource. Contrary to what Hardin thought, they are not necessarily in a system that compels them to defect. Resource users tend to meet frequently (whereas the prisoners met only once and could not communicate with each other) and they have an incentive to keep their engagements for this increases the chances of the fellow users also keeping their promises. If resource users can overcome the free-rider problem then they can also ensure the conservation of jointly used resources.

The arguments of state control and privatisation advocates are also based on the notion that if a resource is not privately or state owned therefore it is not owned at all, it is free for all to use. Not all jointly used resources are, however, open access resources. Many of them are held as common property, which means they are owned by a restricted group of individuals or families, in which case the arguments regarding security of access and well defined property rights will apply as they do for privately owned resources. If a common-pool resource is scarce and under open access conditions, individuals may indeed overuse it, but if the resource is common property many of the arguments of advocates of private property apply, and there is no a priori reason to assume that the resource co-owners will deplete it. Hardin himself acknowledged this in a later article (1991), where he specified that the depletion of jointly used CPRs only occurs if the resource is scarce and if individuals fail to regulate its use. 'Clearly, the background of the resources discussed ... by myself was one of non-management of the commons under conditions of scarcity' (Hardin, 1991, p.178).

Common property is the third potential solution for conserving common-pool resources. Common property, however, has also often been considered the reason why common-pools are depleted. To understand why this is not so – common property sometimes fail to ensure the conservation of CPRs, but it is not in itself the cause of their destruction – it is necessary to be aware of the concept of 'property'. Contrary to what is often assumed, 'private property' is not synonymous with 'property' (MacPherson, 1978), since property can take many different forms, private property being just one of them. Lawyers, economists and political scientists use the term 'property' differently, but they agree on some points. One is that property refers to a social relation with respect to things and resources (Bromley, 1991). This social relation is composed of rights; property refers to rights over things or resources rather than to the things or resources themselves. Property rights can include the right to use the resource, to rent it, to change the look or subsistence of it and to transfer those property rights (Pejovich, 1995). Private property is when all these rights are concentrated in one single owner, but these rights may also be spread out between different owners.

The use of the terms 'property' and 'private property' as synonymous, stems from historical developments that cannot be generalised across time and space. The meaning of property is cultural and time specific and it was only with the rise of capitalism that the concentration of all rights in one individual or corporation became more widespread[2]. As landed property rights began to be transferred all in one block, the term 'property' started to be used to refer to the thing or resource that is the object of these rights (Mac Pherson, 1978). Hence, the term property, especially in the Western world, became an informal synonym of private property.

> Inevitably for us westerners [our notions of property] are rooted in our own particular historical experience. Broadly speaking, our attitudes to property are associated with the development of capitalism and with the notion of commodity. Property for us is based on the idea of 'private ownership' which confers on the individual the right to use and disposal ... these concepts are historically and culturally situated in the western tradition (Hirshon, 1984, p.2).

Property can take a wide range of different forms and '[p]olicy analysts who would recommend a single prescription for commons problems have paid little attention to how diverse institutional arrangements operate in practice' (Ostrom, 1990, p.21-23). Common property institutions refer thus to a combination of common and private rights over something, in this case a resource. Resource users usually have private rights over the resource units and common rights over the resource system, but which rights are common and which rights are private vary from case to case.

Another aspect of 'property' which is generally agreed on is that by definition property rights include security of access to the resource and the existence of duties as well rights (Bromley, 1991). A property right represents a secure claim to use or benefit from the object of property. For a claim to be secure and to be a right, the state, society, custom or law must enforce it. 'A right is the capacity to call upon the collective to stand behind one's claim to a benefit stream. Notice that rights only have effect when there is some authority system that agrees to defend a rights holder's interest in a particular outcome' (Bromley, 1989, p.16). In other words, the authority system must decide who has the right to the benefit stream and who has the duty to respect this right. Therefore, for a right to exist there must always be a correspondent duty. Hence, to be a property rights institution, common property must include some security of access and exclusion of non-owners, who have the duty to respect the rights of the owners to their resource. If everyone has access to the resource, no concept of property is involved.

Issues of terminology have been central in the debate on common property; the term 'common property' has been alternatively used to refer to a type of resource and to an institutional arrangement under which resources are held[3]. As the use of the same term to refer to different phenomena is naturally the source of considerable confusion (Ciriacy-Wantrup and Bishop, 1975; Ostrom, 1986; Bromley, 1991), the present study uses the term common property to refer exclusively to an *institutional arrangement* that involves a restricted group of owners. CPR or common-pool resource refers to the *type of resource* (see Table 1.1).

Table 1.1 Types of goods and property rights institutions

Types of good	Types of property rights institutions
Private good	Private property
Common-pool resource	Common property
Public good	State property

A resource management regime is 'a structure of rights and duties characterising the relationship of individuals to another with respect to that particular environmental resource' (Bromley, 1991, p.22). A common property regime is when 'the management group (the 'owners') has the right to exclude non-members, and non-members have the duty to abide by exclusion. Individual members of the management group (the co-owners) have both rights and duties with respect to the use rights and maintenance of the thing owned' (Bromley, 1991, p.31). A common property regime is thus a socio-economic institution, either legally established or in the form of a customary arrangement, in which non-owners can only use the resource with the owners' permission. Common property regimes often exist in 'state lands', in which case the resource users must comply with state regulations, but nevertheless the co-owners have most of the decision-making rights over the resource; formally, this is the situation of extractive reserves, which belong to the state, but are managed by their inhabitants, rubber tappers and other extractivist populations. Property is rarely absolute; for instance, the private owner of a plot of land may have the right to use and transfer his or her rights to the resource, but lacks the right to build on the area. Property is the right to use a resource 'in any way that is not prohibited' (Reeve, 1986, p.12), and the legal national framework limits both the rights of co-owners' of a CPR and of private owners.

The specifications of the property rights vary from case to case and involve, for example, who has access to the CPR, under which conditions, which rights are individual and which rights are held in common. Whichever way the rights are distributed, the co-owners have the right to use the resource whether they actually use it or leave it rest, and thus, they do not need to extract units from the CPR to establish ownership (Ciriacy-Wantrup and Bishop, 1975). In other words, it is for them unnecessary to extract resources today because they may be unable to do so in the future. The argument of market environmentalists that private property can promote conservation because it provides security of access thus applies to common property as well. (In fact, some market environmentalists consider common property as a form of 'private property'). Although co-owners have the right to extract units from the common-pool resource, they cannot dispose of the resource system because it belongs to the group. The property rights to the resource system are held in common and thus management decisions cannot be taken individually. The co-owners privately own the units that they extracted, and take alone all decisions concerning the resource units extracted, provided that they are in accord with the resource system rules[4].

So far, I have argued that individuals jointly using a CPR can overcome the free-riding problem and set up common property arrangements that can conserve the resource. From this, however, it does not follow that all individuals develop such institutions or that all resources held as common property are safe from destruction. The second part of this chapter develops a framework for examining the conservation of shared resources. Contrary to most other frameworks examining CPRs, the one presented here takes into consideration the fact that resource users rarely live in isolation and thus their context – including government policies and international trends – is included in the analysis. A preliminary outline of this framework can be observed in Figure 1.1.

Figure 1.1 Common property institutions and the conservation of CPRs (I)

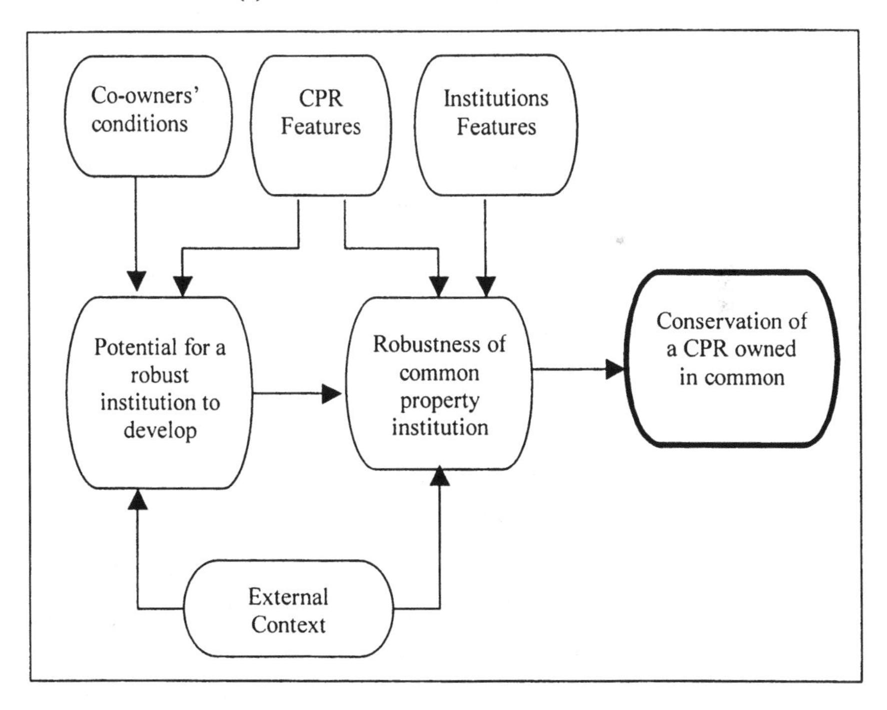

The conservation of a CPR owned in common depends largely on the features of the common property regime, which in turn is contingent on the jointness and exclusion conditions of the CPR. The next section reviews these conditions and proceeds to discuss the features that regimes that have secured the sustainable use of CPRs over time – robust regimes – tend to share. If a common property regime, however, fails to share the 'robustness features', this is no indication that the CPR has to be privatised or put under state control; co-owners of a CPR are unlikely to set up a robust regime until the need for doing so arises, that is, until the resource becomes scarce. After reviewing robust regimes, the chapter turns to examine several factors that influence the potential of a common property regime becoming a robust one. The features of a regime as well as the potential for a robust regime to develop are also contingent on the external context where CPR and co-owners are embedded, an issue examined near the end of the chapter.

Robust common property regimes

A robust regime is one that can ensure the conservation of a shared resource over time. When examining the conservation potential of a common-pool, one must consider the jointness and excludability characteristics of the resource and the institutional arrangements under which it is held (Oakerson, 1986). The excludability features of the resource will influence the easiness with which depletion of the resource by non-owners can be avoided. The depletion of many common resources has occurred because outsiders disrupted the common property arrangements of the co-owners, rather than because of the free-rider problem. This tends to occur when commoners' rights are unprotected. To have secure property rights, however, is insufficient to ensure the conservation of a common-pool. The jointness characteristics of the CPR will influence the difficulties of ensuring that co-owners do not overuse the resource. An additional point to note is that a regime may be able to deal with both jointness and excludability issues for long periods of time, but when conditions change, e.g. outsiders arrive in the area or the state attempts to take over the resource, they cannot ensure the conservation of the CPR any longer. A robust regime is thus more precisely defined as one which is adequate to local circumstances and that can deal with changes in the circumstances. In what follows, robust regimes' characteristics are identified. This is largely based on the groundbreaking work of several scholars who have examined empirical evidence on the success or failure of common property institutions in conserving CPRs over long periods of time (see Ostrom, 1990; 2001; McKean; 1992; 1998; Singh, 1994).

The difficulties arising from the jointness and exclusion conditions of the CPR differ from case to case, and so each group of co-owners will have to devise mechanisms to tackle the specific problems of the resource they use. Depending on the physical features of the CPR – e.g. whether it has well defined boundaries and how large the resource system is – the main difficulty may reside in either harmonising the co-owners' use of the resource or restricting access to the CPR by other individuals or groups. The case of a lake owned by a fishing village can illustrate the first situation: it is easier to exclude people from outside the fishing village from fishing in the lake, than to monitor how much fish is caught by each boat from the village. For inhabitants of a hunting village, on the contrary, it is easier to monitor hunting by each villager, than to control hunting by people from other villages, who live in the distant regions where animals migrate to. A similar type of difficulties exists in the case of a ground basin to which the inhabitants of several villages have access through a number of wells, one in each village. For the inhabitants of each village it may be relatively easy to harmonise their own use of the water basin because they can extract water from one well only, and thus they can monitor the villagers' use of the well. But, they also need to harmonise their use with that of the other villages, which assumedly also have wells – this is likely to be more difficult than harmonising their own use since people from one village cannot monitor how much water is being extracted from wells in other villages. The situation in the Amazon rainforest is very similar: a community of extractivists can ensure the conservation of their own forest with relative ease, but it is extremely difficult for them to prevent other users from misusing the forest.

The degree of jointness of a CPR depends also on the available technology. A fishery, for example, may be jointly and sustainably used while the technology available is not sufficiently powerful to permit large catches of fish. Once resource users have access to new technology, such as refrigerators that permit conservation of fish over long periods, the existing institutional arrangements may cease to be capable of ensuring the conservation of the fishery. The jointness conditions of the resource have changed and this may require the development of new institutional mechanisms to prevent the overuse of the CPR. The excludability conditions of a CPR depend also on existing technology. Before the invention of barbed wire in the late Middle Ages, the excludability conditions of pastures were different from the current ones, and this is reflected in the type of property rights institutions used for pastures then and now – common ownership of pastures was more frequent in medieval times than at present (Oakerson, 1986).

Another factor influencing the exclusion conditions of CPRs is the socio-economic context of the area where the resource is located. For many communities in the Third World, barbed wire is an expensive commodity and thus to divide their pastures is more costly (and hence difficult) than to divide those of herdsmen living in economically developed regions. The existence of a strong police force may also make excludability easier (Oakerson, 1986). Finally, the difficulties involved in securing the sustainable use of a CPR are contingent on the use that is made of the resource (Messerchmidt, 1986). In the case of a forest, collection of fuel wood has a higher degree of jointness than harvesting of timber, which can easily lead to the depletion of the resource. A resource can also be used jointly for different purposes, such as for collecting both fuel wood and forest fruits, but some joint uses are incompatible, such as harvesting timber and tapping rubber.

Apart from addressing the specific difficulties involved in the sustainable use of a common pool resource, a robust regime is one that can cope with a certain level of change in the circumstances. *A change in the circumstances* may refer to natural disasters (e.g. a drought), an increase in the population (e.g. due to an increase in the fertility rates of the commoners), changes in the technology available to the co-owners of the CPR, a change in access to markets, or alterations in the wider legal framework where the resource users are embedded. *A certain level* of change in the circumstances is difficult to define because what can lead to the disruption of a regime in one case may only result in an alteration of the regime rules in another case. For instance, the state frequently disrupts common property regimes and endangers the conservation of common resources because it attempts to take control over jointly used CPR, yet there are many cases of commoners who when faced with state regulations, ignored them and continued following their own rules, or campaigned to obtain state recognition of their management systems. Whether or not a common property regime will secure the conservation of a CPR when changes in the circumstances occur depends on a combination of many different factors, but a 'robust' regime is more likely to tackle these difficulties than a 'weak' regime.

The definition of a robust regime is thus a common property institution that has the necessary mechanisms for preventing the destruction of the

CPR by both outsiders and co-owners, and for coping with a certain level of change in the circumstances. The specific mechanisms of each regime vary according to the characteristics of the resource and the socio-economic context. Nevertheless, studies (see National Research Council, 1986; McCay and Acheson, 1987; Berkes, 1987; Wade, 1987; Ostrom, 1990; 2001; McKean, 1992; 2000; Ostrom, Gardner and Walker, 1994) comparing different regimes that have succeeded in conserving CPRs over the years suggest that these institutions share some general features:

- The boundaries of the resource and of the group of users are well defined.
- There are rules governing the use of the resource and a diversity of mechanisms for securing compliance with the established rules.
- The co-owners of the resource have sufficient autonomy to manage their own resources.

A defining characteristic of common-pool resources is that their boundaries are not clearly defined, hence the problem of excluding potential beneficiaries. It is impossible, however, to own something and enjoy all the advantages of holding property rights without knowing what is owned and who are the owners and the non-owners. To address this problem users of common-pools have devised various ways of defining the boundaries of the resource and of the group of co-owners, the commoners, so that it is clear to the users what they are managing, and who is allowed to use the common-pool. In the case of fisheries, for example, the fish itself is mobile, but many fishing villages have defined the boundaries of the resource by establishing 'fishing territories' (e.g. the area up to 15 km from the coast) where only the members of the group are allowed to fish (Berkes, 1987). The members of the group may be all the inhabitants of the village, or all the males of the village over a certain age, or all the households of the village (each household counting as one member only). Whoever forms the group, this needs to be clearly defined in order to monitor the use of the fishery.

As the joint use of CPRs is non-rival only up to a certain level of use, robust common property institutions usually have rules for harmonising the co-owners' use of the resource. These rules must be specifically designed for conserving the resource (McKean, 1992). All property rights regimes involve a certain degree of specification, e.g. who has access to what, but such rules can have other objectives than the conservation of the resource; the rules can be related to inheritance and social harmonisation instead. In such cases, the conservation of the resource is often due to favourable circumstances, and a change in the circumstances, such as sudden access to markets where to commercialise the resource units, can result in the depletion of the common-pool (Demsetz, 1967; Hames, 1987; Carrier, 1987). Rules, however, can be very complex, and as several anthropologists have pointed out, it is difficult for outsiders (development officials and researchers) to differentiate when the rules aim at protecting the resource and when they have other objectives (Hames, 1987; Carrier, 1987).

McKean (1992), in her examination of a wide range of case studies (contemporary commons in Japan, Middle Ages English Commons, the Swiss mountain commons, and common property institutions in the developing world) highlights some additional features of long-enduring common property regimes. In all the cases she investigated, the environmental rules were clear and highly specific to the local conditions of the resource. In the case of pastures, there were, for example, limits on total size of grazing herds or on the period of time the common grassland was open to use; co-owners of forests regulated the products that could be taken from the commons and the tools that could be used. Commoners also tended to choose rules that were easy to enforce; for instance, instead of setting up a rule stipulating that during a certain period of the year users could only collect one type of product, they preferred to close all access to the commons during that period. They found it easier to monitor entrance into the commons than to verify which products each co-owner collected from the CPR.

In most robust regimes, compliance with the rules is monitored because otherwise individuals may be tempted to free ride on their fellow commoners' conservation efforts (McKean, 1992; Ostrom, 1990). In the case of small communities, social pressure may prevent resource users from taking advantage of each other, but this is unlikely to be the case when groups are large or when given the characteristics of the resource people can defect without being noticed. Large groups are also less likely to be concerned with the interests of the community as a whole, since they may not know the other resource users or may not identify with them. Explicit monitoring of the rules is therefore essential. This, however, can take many forms; what matters is that the commoners themselves are in the final instance responsible for monitoring the CPR, but it is unnecessary that they do the actual monitoring themselves. Some communities prefer to have external agents (appointed guards and detectives) monitoring the rules rather than controlling each other's actions, but these agents are accountable to the co-owners of the resource (Martin, 1979). Alternatively, monitoring mechanisms can be embedded in the rules, as in the case of closure of the commons referred above.

In case of defection, there is a penalty: monetary penalties (Ostrom, 1987; Arnold and Cambell, 1986), destruction of equipment (Acheson, 1987), or withdrawal of access rights (Easter and Palanisami, 1986). Successful regimes usually have gradual penalties (Ostrom, Gardner and Walker, 1994). The application of a small penalty for a user who breaks a rule for the first time will be sufficient to remind the co-owner that the regime mechanisms to prevent free-riding are efficient; this in turn increases reliance on the system and, by extension, compliance with the established rules. On the contrary, imposing a large penalty may create resentment and unwillingness to conform with the rules in the future.

The existence of rules and enforcement mechanisms is, however, insufficient to prevent conflicts arising among users. Users can interpret the rules differently, which in turn can give rise to free riding. The following example can illustrate this problem: to conserve an irrigation system co-owners agree that each household should contribute one individual to work one day in some common task (e.g. cleaning the canals of an irrigation system); however, some households start sending children and old people to do this work. If all other households send a working adult, the family sending the child or old person is free riding on the other commoners' efforts. The remaining families will have less interest in contributing to the common work if they see that they are carrying a higher share of the cost involved in providing the common good than the free-rider family. This may lead them to do the same; but if all families send their less productive members to do the common work, the rule 'each family must send one member to do one day's work' may become insufficient to secure the conservation of the CPR. To prevent that this type of problem undermines a common property institutions it is necessary to have a system to solve conflicts and decide on which interpretation of the rules is correct. Robust regimes tend to have low-cost conflict resolution mechanisms, such as arenas for solving conflicting interpretation of rules, and decision-making systems for dealing with these issues, such as an assembly of village elders (Ostrom, 1990).

Finally, in most robust regimes, individuals have sufficient autonomy to manage their own resources and their rights to manage the common-pool are recognised by the state (see Berkes, 1986; Thomson et al, 1986; Easter and Planisami, 1986; Bromley and Cernea, 1989; Lawry, 1990; Chopra et al, 1990; McKean, 1992; Hilton, 1992). If the co-owners lack autonomy to manage their CPR, it will be difficult (but not impossible) for the group to enforce compliance with the rules by its members. To restrict outsiders from using the resource may be even more difficult if the group common ownership of the CPR is not legally recognised and the law supports people outside the community using the resource.

Table 1.2 summarises the features of robust regimes. A common property institution is considered robust when the boundaries of the common resource and of the group of resource users are well defined, the co-owners have exclusive rights, there are rules specifically designed for ensuring the conservation of the resource and enforcement mechanisms to secure compliance with these rules. In addition, resource users should have sufficient autonomy to manage their CPR and their rights supported by the state.

Table 1.2 Features of robust regimes

Boundary rules	**Harmonisation rules**	**Monitoring and Enforcement mechanisms**	**Autonomy**
• CPR boundaries well defined • Group of co-owners well defined	• Aimed at conservation • Locale specific • Clear • Easy to monitor	• Co-owners responsible for monitoring • Gradual penalties • Conflict resolution mechanisms	• Exclusive rights of co-owners respected • Co-owners' rights to manage CPR respected

An institution meeting all these requirements is more likely to secure the conservation of the common-pool in the long-term than one where, for example, co-owners have exclusive rights to the resource but fail to regulate their own use of it. In the latter case, a change in technology or market prices may lead co-owners to extract too much from the resource system whereas in the case of a robust regime commoners will probably tighten their rules to ensure that extraction rates do not endanger the common resource. The chapter now turns to review the conditions usually present when individuals develop robust common property regimes.

The development of robust common property regimes

For commoners to develop an institutional arrangement to conserve their resources, it is necessary first that they perceive the need for such an arrangement. In most cases, the limitations of joint use of a CPR only become apparent once the resource is used at its limit and, consequently, the necessity to develop institutions to conserve a common-pool arises only if the resource becomes scarce. If demand for the resource units is low and does not exceed the supply or regenerative capacity of the resource system, an open access common-pool can be sustainably used. (This may be the case when the group of potential users is very small or when the available technology imposes restrictions on how many resource units they can extract.) If the demand however increases, due to new technology, to an increase in the size of the group of users, or to an increase in the market value of the resource, the common pool may then become scarce making it necessary to establish mechanisms that limit the extraction of units from the common-pool by all potential users (Bromley and Cernea, 1989; Easter and Palanisami, 1986).

Then again, and as Lawry (1990) notes, scarcity of the resource will not necessarily lead to the establishment of common property arrangements: 'Resource scarcity can lead to co-operative management or grater competition and individual action to privatise the resource' (1990, p.412). In East Africa, for example, there are enclosure movements taking place on rangelands partly because of increased scarcity of resources. The Tigray in Ethiopia alternate between almost-private property when population increases (and therefore the CPR becomes scarce), and common property when demand for the common lands is less strong (Bauer, 1987). Scarcity of the resource is thus one of the factors usually present when common property institutions develop, but it is not a determinant factor.

The more dependent commoners are on their CPRs, the higher will be their need to conserve them. From this follows that individuals whose livelihood is largely based on shared natural resources are likely to develop mechanisms that prevent their destruction (see Easter and Palanisami, 1986; Bromley and Cernea, 1989; Ostrom, 1990; Lawry, 1990). A study of villages in the south of India by Wade (1986), for instance, showed that the villagers tended to develop common property arrangements for the management of water to meet intensively felt needs that could only be met through co-operation. The examination of Indian common lands by Chopra et al (1990) showed similar results. In general, regulations for securing the conservation of the resource were established for those resources that were especially important for the community, such as water in arid and semi-arid environments and forests in situations where people obtained their living from them.

Second, apart from the need to conserve the CPR, whether or not commoners will set up a common property institution depends on the potential of the resource for being managed under such an arrangement. That is, a resource may be scarce and users highly dependent on the common-pool but it may not be possible to exclude outsiders or even monitor co-owners' extraction rates. Hence, the potential for setting up a regime varies in relation to the clarity of the resource boundaries and the excludability conditions of the CPR (Wade, 1987; Bromley and Cernea, 1989; Lawry, 1990; Ostrom, Garden and Walker, 1994). For example, given the same levels of scarcity and dependency on the CPR, pastoralists are more likely to set up a robust common property regime than fishermen are because a pasture has clearer boundaries and exclusion of non-owners is easier[5]. As the excludability conditions of common-pools are dependent on the available technology, technological developments also influence whether resource users decide on private or common property or leave the CPR under open access. If individual exclusion is easy, private property may be the preferred solution; if the cost of exclusion technology (such as fencing) is very high, a common property might be better (Wade, 1987); and if exclusion is quasi-impossible (as it is the case with most global CPRs such as the atmosphere), resource users may leave the CPR under open access.

A third set of factors that influences the probability of users of a CPR setting up a common property regime concerns the users' capacity to establish such a regime. Their capacity for developing rules for conserving the CPR will depend on their knowledge of the resource, on their access to information, on the size of the group and the homogeneity of the members, and on their autonomy from higher political bodies, such as the national state.

One of the advantages of common property over state property that scholars usually point out is that users of a common-pool have an intimate knowledge of the resource whereas state technicians have to acquire this knowledge before being able to design rules that ensure the sustainable use of the CPR. Due to their long-term use of the resource, local communities have in-depth knowledge of their natural resources. But with regard to unexpected changes in the quality of the resource, of which there is no past experience, some commoners may explain them in mythological terms, in which case they may take to long in establishing the relationship between human use and resource depletion and fail to prevent the deterioration of the common resource (Hames, 1987). The higher the ecological knowledge of the resource users, the higher the chances that they will set up a robust common property regime, and, as Wade (1987), comparing various successful regimes, points out: 'the better their knowledge of sustainable yields the greater the chances of success' (1987, p.104).

Apart from information on the ecological features of the CPR, to begin the process of establishing a common property regime, resource users need also information on the expected costs and benefits of the proposed change (Ostrom, 1990). Information on the range of opportunities that may or may not be available to them outside a particular situation will influence their decisions, as well as the users' knowledge of the norms shared by other relevant actors. Which information the users of the resource have access to, how they obtain information, whether their information is biased or not will all affect their decisions. The availability of information depends in turn on a number of interrelated factors, such as the size of the community, the complexity of the CPR and the support users of a common-pool may receive from outside specialists.

It is generally agreed in the literature that the smaller and more homogenous the group of resource users is, the more likely it is that they will endogenously set up the necessary mechanisms for conserving their common-pool (Wade, 1987; Bromley and Cernea, 1989; Lawry, 1990). Large and heterogeneous groups, however, can also develop robust common property regimes. This issue – the relevance of size of group of commoners in the development of robust common property regimes – is particularly important in the study of extractive reserves. Many reserves are inhabited by more than 300 families, and the Chico Mendes Reserve has nearly 1,500 families – a small group has at most 300 members (Ostrom, 2001), and according to other sources, only150 members (Hardin, 1991).

When discussing the limitations of collective action, Olson (1971, p.50) considers that the 'common good' can be provided if the group 'does not have so many members that no one member will notice whether any other member is or is not helping to provide the collective good'. Hardin, in his 1991 revisionist article, shares this opinion and argues that small communities of up to 150 people can manage their resources and that free-riding incentives are checked by 'shame', whose effectiveness depends in turn on face-to-face confrontations. This argument has been further elaborated Singleton and Taylor (1992) who argue that only those groups that form 'communities' can develop endogenous solutions for the conservation of CPRs. They define 'community' as 'a set of people (i) with some shared beliefs, including normative beliefs, and preferences, beyond those constituting their collective action problem, (ii) with a more or less stable set of members, (iii) who expect to continue interacting with one another for some time to come, and (iv) whose relations are direct (unmediated by third parties) and multiplex' (p.315). In addition, the members of the group should be 'mutually vulnerable actors', in the sense that 'each of whom values something which can be contributed or withheld by others in the group and can therefore be used as a sanction against that actor' (p.315).

Ostrom (1992), however, argues that the existence of a community is an important factor in the solution of CPR problems, but that it is neither a sufficient nor a necessary condition for the development of common property arrangements. She has strongly refuted the claim that only small communities can endogenously manage their common-pool resources, and supports her argument with a number of well-researched examples. One is the case of some agricultural communities in Spain, which were unable to prevent the depletion of their irrigation systems although they met all community features described by Singleton and Taylor (1992). Users of underground water basins in California, on the other hand, established a common property arrangement to avoid the depletion of their water supply although they were numerous (800 – 12 000 users) and did not meet any of the community characteristics, such as homogeneity, described by Singleton and Taylor (1992).

The Californian commoners do not represent an isolated example of large and heterogeneous groups developing mechanisms to conserve their common resources (see Agrawal, 2000). Research by Gadgil and Iyer (1989) shows how seven villages in India shared and managed their natural resources for a long period although the villagers formed a large group and there were strong differences among them because of the Indian caste society. Social relationships among the different castes were virtually inexistent (e.g. there was no inter-marriage), and some resources were secured for one group only. All the same, some resources could be used by all groups, and there were rules for the management of all resources as one whole – this common property regime was sustainable until some breakdowns occurred resulting from the centralisation process of British colonial rule[6]. Hilton (1992) also examines how large groups have set up common property arrangements to manage their resources. She presents the case of two irrigation systems, one shared by seven villages and the other by five villages. The inhabitants of each village do not necessarily communicate with those of the other villages sharing the same irrigation system, but discussions to set up arrangements for using the systems take place between the village representatives. These two irrigation systems are successful in the sense that there is a high level of community participation in the conservation of the CPR, which has been well maintained.

Large groups – assumedly using large and hence complex resource systems – tend to develop common property institutions in the form of 'nested enterprises' (Ostrom, 1990). When the resource is complex, its sustainable use may require rules at different levels, but these rules need to be harmonised. The case of a groundwater basin used by people from two villages may illustrate how a system of nested enterprises works. People from the first village take water from a well located in the market place, people from the second village take water from several wells scattered outside the village. To ensure that the basin is not exhausted, it is necessary to monitor the use of water by everybody, yet each village needs different rules and monitoring mechanisms. To ensure that people from the first village take only the allocated amount of water, it is only necessary to check that no one takes more than one bucket of water from the well, and as the well is located in a public place, water extraction in the first village is relatively easy to monitor. To ensure that people from the second village take only the allocated amount of water it is necessary to check how many buckets they bring when they come back from collecting water, since it would be very costly to have one monitor next to each well. In addition, the rules of both villages need to be harmonised, since it can happen that people from one village take more water than do people from the other village. The mechanisms that villages need to monitor each other's behaviour are again different from those that each village uses to check on its own inhabitants' water extraction levels.

For a common property regime to develop, the most relevant feature of the resource users is thus not the size of their group. Moreover, for a complex CPRs such as the one described above, the development of a robust common property regime may in fact require a rather large group of co-owners. Given that all other factors are equal, users of a CPR are most likely to develop a robust regime if they share generalised norms of reciprocity and trust, but although small groups may be more likely to meet this requirement than large groups, the latter may also develop relationships of trust. In some cases, norms of reciprocity and trust may develop out of the common endeavour to develop a robust regime, provided that all users perceive the need to manage their CPR and all other factors for the development of common property arrangements are met (Ostrom 1990; 1992).

Lastly, the development of common property regimes depends on the level of autonomy resource users have. In the case of jointly used CPRs which are managed by the state, direct users may ignore the need to take an initiative to reverse a potential depletion process because they see this as the responsibility of the state (Ostrom, 1990). Developing institutional arrangements to secure the conservation of a CPR involves considerable effort. Resource users must take time to discuss the problem, find information on what they can do to solve the problem, define rules and monitoring mechanisms and finally, enforce compliance with the rules established. If they believe that an external agency can provide the common good, which in this case is the conservation of the CPR, they are likely to leave the state to carry the costs of it. The shortcomings of state control of CPR have already been reviewed. Autonomy, like all the other factors conducive to the development of robust regimes, is not, however, a *sine qua non* condition. If resource users consider that state management is ineffective in ensuring the conservation of the CPR, they may engage in action to develop institutional arrangements in spite of state control or even in direct opposition to it (Cordell and McKean, 1986; Acheson, 1987). Different groups fishing lobster in the USA, for instance, prevented access to their fisheries to any potential fisherman who was not a member of their group although legally anyone who had a licence to fish could use the fishery (Acheson, 1987).

To summarise, the development of robust common property arrangements depends on the interaction of several factors (see Table 1.3).

Table 1.3 Factors influencing the development of robust regimes

Need for regime	Characteristics of the CPR	Characteristics of the users	Autonomy
• Perceiving need	• Jointness conditions	• Small group	• Autonomy to set up their own management rules
• Scarcity of the CPR	• Exclusion conditions	• Homogeneous group	• Autonomy to enforce their own rules
• Dependency on CPR		• Knowledge of CPR	
		• Information	

Individuals only develop institutional arrangements to ensure the conservation of their common resources when they perceive the need for it, which usually occurs when the CPR becomes scarce. If they are highly dependent on the resource, their need to conserve it is naturally high and hence their incentive to engage in its conservation is likely to be powerful. Yet, common property will only develop if, given the jointness and exclusion conditions of the CPR, it is a feasible solution and more advantageous than alternative property rights institutions. Resource users who have sufficient autonomy to manage their resources are also more likely to develop robust regimes than those whose autonomy is hindered by a controller state. Finally, if all other factors are similar, a small group of resource users, whose members are homogenous and have direct and multiplex relationships, and who expect to continue interacting over time are more likely to develop a robust regime than groups who meet none of these features. If the resource is large and complex, however, a small group who has control over only part of the CPR will be unable to secure the sustainable use of the resource. Large groups whose members are heterogeneous can also develop robust regimes, in which case the regime may take the form of nested enterprises.

So far, we have shown how the characteristics of the resources, their users and their users' regimes influence the sustainable use of a common-pool. During this discussion it became apparent that the context in which the resource users are embedded also plays a role: the level of autonomy commoners have, for example, depends on the state policies of their country; also, resource users rarely develop the technological innovations that later impose radical changes in their lives. The influence of the external context in the development and robustness of common property regimes is not, however, limited to these two factors, as will be discussed in the next section.

The external context

In general, the role of the external context in affecting the choices of commoners has received less attention than the internal factors mentioned in the previous sections. However, 'in a globalizing world, ecological, economic, social and political interdependencies have but reinforced the impact of external factors on socio-cultural entities defined as geographically bounded wholes' (van Ginkel, 1998, p.2). A recent paper by Edwards and Steins (1998) has defined contextual factors remote from the CPR as those that 'have an indirect influence on the situational variables of the CPR and are usually outside the control of the user community' (Edwards and Steins, 1998, p.4). These factors will affect what is politically, physically, economically and socially feasible in the context of CPR management (Edwards and Steins, 1998). The external context however also has direct effects on CPR users, e.g. government policies which support the colonisation of areas held under common property, will result in a direct confrontation between the co-owners of the resource and the newcomers. The term 'external context' as used here thus refers to all those issues that may influence the development and resilience of a common property institution, but that do not result directly from either the physical characteristics of the resource, or from the characteristics of the group of users, and over which users of CPR lack complete control, although depending on the circumstances they may have some.

Including the external context in the examination of a common property regime poses a difficulty, which is finding a theoretical framework to analyse the role of external factors in the conservation (or depletion) of jointly used CPRs. Within the theory on common property regimes, there are scarce analytical tools to examine the external context. Existing frameworks address almost exclusively internal factors (Oakerson, 1986; Ostrom, 1990; Singh, 1994) and about the only external factor considered is the role of the state and the autonomy it leaves to the commoners.

The literature on 'environmental action' and local-level participation in resource management (e.g. Ghai and Vivian, 1992; Friedmann and Rangan, 1993; Collinson, 1996) also addresses issues relating to common property institutions. This body of research tends to pay considerably more attention to the external context than the theory on institutional choice; however, scholars in the area tend to ask a different set of questions from those examined in this study. My subject of inquiry is the influence that external factors may have on the institutional arrangements to conserve a common resource. By contrast, writings on environmental action focus on the political and social aspects of sustainability and common property regimes, and examine external factors in relation to issues of power, social justice and conflict, rather than their influence on the capacity of the commoners' institutions to ensure the conservation of the CPR. This body of literature is, however, very useful to identify some of the factors that may influence commoners and their sustainable use of common resources.

Given the absence of a well-defined theoretical framework for examining the external context, it is not easy to define which external factors we should include in the analysis[7]. Van Ginkel (1998) argues that in a 'minimal framework, attention should be devoted to ecological, demographic, infrastructural, technological, economic, political, legal, social, cultural and religious factors impinging from the external world on localised systems of common pool resource use and the adaptive responses of the users (van Ginkel, 1998, p.11). The problem with this approach, however, as Van Ginkel himself recognises (p.2), is how to set the boundaries to the external context: 'we should take account that the blurring of boundaries is part of the problem we are dealing with'. Edwards and Steins (1998, p.8) acknowledge the same problem: 'Clearly, there is a limited extent to which the researchers of a specific resource system can analyse the 'external world' of the common in terms of contextual factors'. To identify the external factors that other researchers have most often referred to as having an impact on the conservation of jointly used CPRs can partly help to address this difficulty. Taking this approach as a starting point, the remainder of the chapter reviews the principal factors that an analysis of a commons' external context of should look at.

External factors can affect the conservation of a jointly used resource in two ways: they can provoke a change in the circumstances and they can influence the capacity of the resource co-owners to deal with any such change. A change in the circumstances can provoke three types of outcome. First, alternative property rights institutions can develop; given the change in circumstances, private or state property may become more appropriate to the use and conservation of the CPR than common property. Some of the Japanese commons examined by McKean (1992) are a case in point. Although successful for long periods, they eventually became uneconomical because of changes that occurred in the external context, leading to their disappearance[8]. The villagers had used the commons to obtain fodder and fertiliser for their private fields but now, by selling their crops, they could buy these products from the market, and this was a better option than to continue working in the commons. Hence, many villagers sold their commons. Second, resource users may adapt their regime's rules and enforcement mechanisms to the change in circumstances and in this way secure the conservation of the CPR. This is more likely to occur if the regime is robust than if it is weak, however, a change in the circumstances may also trigger the development of a robust regime. Third, a change in the circumstances may lead to a situation of open access and consequently natural resources might be depleted. The last two outcomes are the ones that concern us here.

The four most frequently mentioned changes in circumstances provoked by external developments are:

- Arrival of outsiders in the area of the commons.
- Changes in the demand for the resource units.
- Increased access to a market economy.
- Changes in government policies.

Commons are frequently destroyed because outsiders (non-owners) take over the resource (see Thomson et al, 1986; Cruz, 1986; Pinkerton, 1987; Goodland, Ledec and Webb, 1989; Bandyopadhyay, 1992). As more individuals compete for the use of the common-pool, the resource becomes scarce. If the boundary rules are not effective, the arrival of outsiders may lead to the development of an open access situation, in which neither co-owners nor new comers have secure access to the CPR. A number of factors can provoke the arrival of outsiders into a commons: construction of a road facilitating access to the CPR, state policies encouraging migration into the area of the commons or increasing demand for the resource units. In most cases, however, threats to the boundaries of a CPR result from a combination of factors. Blauert and Guidi (1992) write about two local communities in Mexico; the boundary rules of their common-pools were threatened because the implementation of irrigation projects, granting of logging concessions and colonisation programmes attracted outsiders to their areas. A community of fishermen in the Indian state of Kerala had also problems with outsiders who had had been attracted to their fishery due to changes in fishing technology, increased international demand for prawns and new state incentives for fishing (Kurien, 1992).

A change in the demand for the resource units can change the commoners' free-riding incentives and if the enforcement mechanisms are not effective, the resource co-owners themselves may deplete the CPR (Cruz, 1986; Durrenberger and Palson, 1987). If demand for the resource units increases, their price will increase too; individual resource users may consider the option of selling as much as possible of the resource units today while the price is high and this will act as an incentive to defect on the established rules of the regime. Demand for the resource units can increase because of changes in the international markets provoked by a change in consumer tastes or by a decrease in the supply of the product from other sources. A decrease in demand for the product with a correspondent lower price can equally be an incentive for resource users to extract more units, to obtain the same income. Changes in the external context can also trigger demand for a new product: a resource unit that was not exploited before and for which the regime has no mechanisms to control extraction. As explained earlier, the jointness conditions of a common-pool resource depend on the use that is made of the resource system; if demand for a new product develops, the jointness conditions of the CPR can thus change and the existing harmonisation rules cease to be able to secure its sustainable use.

Many robust common property regimes exist in the context of market economies (Ostrom, 1990; Berkes, 1987; McElwee, 1994). However, if people living in a subsistence economy have suddenly access to a market economy, this can threaten the conservation of their common resources. Commerce is often an incentive to extract more from natural resource systems in general. Besides, the market economy promotes the extraction of one product only whereas many common property regimes are based on the use of several products, an arrangement that is in general more sustainable (Goodland, Ledec and Webb, 1989). If the rules for using the commons are in the form of traditional values, the development of a market economy may lead to resource depletion because, as several examples have shown, market values tend to be stronger than the former (Goodland, Ledec and Webb, 1989). Alternatively, if the existing rules were specifically designed for securing the conservation of the shared resources, commoners are likely to adapt them to a market economy. The relationship of the rubber tappers with the market economy is, however, different from both situations described above: they were always part of a market economy – rubber prices are determined in the global markets – but their common property institutions were for a long time similar to those of commoners living in a subsistence economy.

Government policies can trigger changes in the circumstances by provoking any of the above-mentioned changes. The arrival of outsiders in a commonly owned area, for instance, is often triggered by national government migration policies. A study of Latin American forests by Richards (1997) shows that state policies are strong causal factors of commons' depletion. After observing that several groups of commoners in the region had successfully adapted to the market economy, new technology and population pressures, Richards (1997) concludes that the breakdown of common property regimes in Latin America is contingent on the role of state policies rather than simply on the effect of any of the other pressures. Apart from centralisation policies, the state has contributed to the disruption of these regimes through tacit or open encouragement of colonisation of areas held under informal common property regimes, policies supporting vested interests groups in the same or contingent areas and failure to uphold basic law and order.

The state also alters the commoners' circumstances when it attempts to take over the management of their shared resources. A community may have a robust system in place, but once their right to manage their resources is removed, it will be difficult for the commoners to enforce the rules they had designed. In Morocco, several tribes had sustainably managed their common lands for centuries until the state, in the 1950s, took control over them (Artz et al, 1986). The state, however, lacked the necessary power to monitor sustainable grazing in the pastures and, as the traditional management regimes had been undermined, the pastures started to deteriorate. Cruz (1986) describes a similar case in relation to a traditional fishery managed by local communities. Once the state nationalised the fishery, the CPR began to suffer. The problem with this system is that while the apparent control of the fishery resource use is assigned to government policy-makers, 'the *de facto* system of exploitation of offshore fisheries is closer to open access. The coastal fishery rules are difficult to enforce with the limited resources available to most municipalities' (Cruz, 1986, p.130).

By definition, robust regimes are more likely than weak regimes to deal with changes in the external circumstances, but such changes can also trigger the development of robust regimes. Threats to people's livelihoods often prompt concerted community action (Blauert and Guidi, 1992). A community may have no boundary rules because the resource has always been abundant. Likewise, their regimes may be weak because technological limitations or existing rules regarding other matters (e.g. inheritance rules) have so far secured the conservation of the resource. However, a change in demand for the product may change this situation and users will need to develop mechanisms to deal with the new threats. Whether resource users will develop robust regimes or adapt the institutional arrangements of their robust regimes to changes in the circumstances, depends not only on the factors described in the previous two sections, but also on the characteristics of the external context where they are embedded.

Several scholars have drawn attention to the influence of the socio-political setting on people's capacity to deal with changes. Egger and Majeres (1992), for example, argue that the capacity of communities to deal with outsiders who try to take over their resources is affected by the overall political setting of each country, which determines the ability of the poor to organise freely. Lack of political rights, such as the right to organise, and restricted access to the media can curtail communities' range of options when devising strategies for regaining control over their resources. When examining the 10-year process of establishing a co-management arrangement between Amerindian tribes and the US federal government, Pinkerton (1992) also stresses the influence of the political context. These tribes faced two problems: outsiders taking over their common resources and an incipient process of resource depletion. Their common resources had well defined boundaries, but lack of legal recognition of their rights made them vulnerable to other agents. In addition, they needed help to secure their own sustainable use of the common-pool in a changed environment. Depending on the overall political climate, the Amerindian tribes resorted alternatively to the courts or to political strategies.

The economic context at any particular point in time also influences the capacity of resource users to deal with a change in circumstances. For instance, given an endogenously triggered increase in population, a community may more easily solve the problem by promoting migration if there are employment possibilities outside the commons. In a favourable economic context (all other factors such as public interest being equal), communities are also more likely to obtain financial support for conservation activities. The socio-economic and political context influences the capacity to attract public interest, which can play an important role in commoners' struggles against the state or outsiders attempting to take over their resources. General public interest in the plight of the Amerindian tribes helped them to obtain recognition and support for their management system (Pinkerton, 1992). A well-known illustration of how public support can make a difference is the case of the Chipko movement in India, whose (initial) success in obtaining control over their forests was largely due to western public interest in their plight, which was in turn related to the place environmental and Indian rights issues had in the concerns of westerners (Bandyopadhyay, 1992; Rangan, 1993). The situation of the rubber tappers was quite similar to this. As it will be shown in Chapter 3, their movement against outsiders was successful thanks partly to the support of the international public, which was highly concerned with the conservation of the Amazon rainforest.

Establishing alliances with specific external agents other than the state (and often in opposition to the state) has frequently helped commoners to successfully address changes in the circumstances (Vivian, 1992; Friedmann and Rangan, 1993). Access to free media and the possibility of voicing their needs helps finding external agents with complementary resources and whose interests are in tune with those of resource users: NGOs, environmental activists, members of a Church, unions and universities. McDonald (1993), for example, describes how local communities in Brazil, in their struggle against the construction of a dam in the Uruguay River Basin, counted on the help of different Churches, rural unions and universities. Likewise, the help of environmental activists and intellectuals greatly advanced the Chipko cause (Bandyopadhyay, 1992; Rangan, 1993). External actors can help with information (e.g. on self-organisation), provide leadership training and facilitate access to media and international organisations. They also can offer advice on political tactics, help with coalition building, and financial assistance (Friedmann and Rangan, 1993).

The legal setting also plays an important role in enhancing or hindering the capacity of individuals to address changes in the circumstances. As mentioned earlier, a robust regime as well as resource-users capacity to develop such a regime is affected by whether common property rights are legally recognised (Ostrom, 1990). Co-owners may have autonomy to manage their own resources, but their right to do so may not be enshrined in the law. There is a large range of other legal stipulations affecting each particular commons. Laws concerning property rights to mineral deposits, for example, can affect herdsmen sharing a pasture. Likewise, laws on logging and environmental conservation partially shape the options that co-owners of a forest have when dealing with a change of circumstances. If legal mechanisms support their sustainable use of the forest, it will be easier for them to develop new regulations to conserve their resources.

There is a debate on the role of the state in relation to common property regimes. As a backlash to the earlier view that all commons had to be managed by the state and given the disastrous results this policy often had, the literature on common property tends to emphasise the importance of commoners having autonomy to manage their resources instead of their need for state support. In recent years, however, it has been pointed out that state support does not imply state management of the resource: 'government intervention does not ... necessarily imply direct action: it may include such indirect measures as enactment of necessary legislation, provision of funds, technical information; guidance and training, establishment of new institutions and organisations, creation of basic infrastructure, etc' (Singh, 1994, p.312). Commoners often need such support and, besides, the responsibility of conserving natural resources should not be left to local communities alone (Yadav et al, 1998). Lawry (1990, p.403) argues that 'while state management is ineffective, incentives for individuals to participate in local management activities are weak, and local institutions are usually unable to generate sufficient sanctions locally to enforce rules'. The abundant evidence of successful common property regimes does not wholly support this statement, but the argument that although the state should not manage commonly owned resources local communities need external help to secure their conservation is pertinent, and will be further discussed in Chapter 7 in relation to extractivists in Amazonia. The stance taken is that the state can either enhance or hinder resource-users' capacity to deal with changes in the circumstances.

Broadly speaking, the state can support common property regimes in four ways. First, it can help commoners secure their resource boundaries against outsiders. Common property regimes, like property rights in general, need the state to secure their rights against other potential users. A facilitative legal framework that recognises common property rights enhances individuals' capacity to deal with other potential users of their common resources. Notwithstanding their legal rights, property rights' holders also need the state to actively protect their rights. The existence of a police force preventing violation of property rights and a judicial system punishing any such violations is crucial, otherwise the existence of legal rights, whilst important, will not be effective in securing the rights. (Chapter 3 shows in more detail the distinction between the three scenarios: having no legal rights, having legal rights but no active protection, and having both.) Second, the state can help individuals to harmonise their use of shared resources by providing access to arenas for conflict resolution. The Japanese commoners mentioned earlier, whose regimes and resources lasted for long periods, had, for example, access to state courts to solve disputes (McKean, 1992). Let us consider the example of a groundwater basin close to depletion and held as common property by several villages. If the villages have access to neutral arenas for discussion, e.g. courts set up by the state, it will be easier for them to initiate discussions to change the rules of their regime than if first they have to set up a framework for having those discussions. For the same reasons, the provision of arenas for conflict resolution can help robust common property regimes to develop. The users of the Californian underground basins examined by Ostrom (1990) resorted to state tribunals to solve their problems. This resulted in the creation of a robust common property regime. Third, the state can provide information to the resource-users. In the example of the groundwater basin, users will more easily identify the limits of their resource if in addition to their own knowledge of it they have access to independent and specialists' studies on their water basin. The Californian commoners co-operated with state organisations in obtaining detailed information on the characteristics of the basins. Fourth, the state can promote the conservation of jointly used resources by providing technical and advisory support to the resource users: small scale infrastructure, new processing technologies, data collection and monitoring procedures, mechanisms for improving communication between users (e.g. radios) and advice on how to build an administrative capacity (Peters, 1986).

State agencies (and other actors) can engage in co-management with the direct users of the resource. By co-management is meant more than only leaving communities sufficient autonomy to manage their resources, recognising their property rights institutions, and providing them with information and arenas for decision making and conflict resolution. Co-management means that communities and the state together manage the resource or some project for the improvement of the users' livelihood or common-pool. 'Co-management can be generally defined as power-sharing in the exercise of resource management between a government agency and a community or organisation of stakeholders' (Pinkerton, 1992, p. 331).

Co-management is suitable when the resource users cannot undertake all the necessary tasks to secure the conservation of their resources. This can occur, for example, if the common-pool is close to depletion and restricted use by the commoners insufficient to reverse the process. Co-management can also help to overcome many problems faced by users of CPRs whose jointness and exclusion conditions make the development of a robust regime difficult. If the resource is very large and its boundaries difficult to monitor, or if there are too many users with little communication channels between them, a co-management system can be more effective in securing the resource than a simple common property regime. Co-management can also be the best alternative when the conservation of a common-pool depends on factors that are beyond the jurisdiction of the community of direct users, which is, to a certain extent, the case with the rubber tappers. The example of a group of lobster fishermen in Maine illustrates this point. Besides legislation to regulate the national lobster fishery, the Maine fishermen had their own rules to secure their common-pool resource (Acheson, 1989). However, developments in the industry and in the legislation caused an increase in fishing, threatening the conservation of the fishery. This prompted one group of fishers to lobby for stronger regulation from the state, for this permitted to limit both their own fishing and that of other groups whose fishing they could not contain. Langdon (1984) reported a similar case in Alaska, where an Aleut community not only controls access to territories within its local lagoon fishery, but also made the state limit fishing in adjacent territories, where outsiders might intercept the local stocks.

Although the state can engage in co-management with users of a common-pool, a facilitative state does not take over the control of the resource; the resource-users keep their common property rights (Ostrom, 1990). In the cases of co-management described above the responsibility for the design of rules and for monitoring compliance rests in the hands of the direct users of the resource. A controller state is one that attempts to manage common-pool resources through state agencies, whereas a facilitative state is one which leaves the responsibility of managing the resource to its co-owners yet supports their common rights in a number of ways. Research has shown that the latter approach gives better results. Hilton (1992), for example, compared a set of irrigation systems that had received direct help with another set that had received indirect external help, concluding that where users had a higher degree of responsibility, the irrigation system had been better maintained and co-owners contributed more to its conservation.

When examining the sustainable use of shared resources it is therefore necessary to look at the external context as much as at internal factors. To identify the exclusion and jointness conditions of a CPR it is indispensable to look at the context in which the resource is located. The external context can trigger changes in the circumstances to which the resource users must adapt: new government policies or alterations in the socio-political, economic, and legal settings can lead to the arrival of outsiders in a common area, to changes in the demand for resource units, to sudden access to a market economy, or to the state taking control over the resource. External factors can also hinder or enhance the capacity of resource users to deal with changes in the circumstances (triggered either internally or externally), even if they have developed robust regimes. More supportive laws or direct help by external agents (including the state) can help resource users to address their problems. The state, for example, can promote the conservation of commons by securing the co-owners' rights against outsiders, and by providing an institutional framework that facilitates discussion and conflict resolution among resource users (especially when they form large groups). It can also provide information and material help. If resource users form a large group or part of the resource is beyond their jurisdiction, the state or other agencies can engage in a co-management regime. Finally, it is necessary to note how the external context affects boundary and harmonisation rules. Whether resource users will need to pay guards to protect their resource is influenced by the effectiveness of the local police force, and whether harmonisation rules need to specify which products can be taken from the common-pool will by influenced by the existence of an external market for a range of different resource units.

Conclusion

This chapter has developed a framework for analysing the use of common-pool resources that accounts for the external context in which resource-users are embedded. The basic argument is that, as many studies have shown, individuals can sustainably use resources in common, but that whether they will do so depends on the wider socio-economic, legal and political setting, as well as on their characteristics as a group and the features of their resource. The theory on common property, which includes different theoretical traditions, from new institutional economics to political ecologists, suggests that the conservation of a commons depends on the regime under which resources are held (which varies from open access to common property), and this in turn depends on the characteristics of the common-pool and its users. External factors are noted as important, but studies focus on how external shocks have disrupted traditional regimes and tend to disregard the more complex influence the external context may have. Besides, 'we need to study common-property regimes from a more dynamic perspective. Not just situations where new 'outside' forces have wiped out local natural resource commons, but more in-depth cases of how particular commons governance procedures were adapted, or not, to changing situations' (Field, 1990, p. 251). This study does not pretend to be the first one addressing the interaction of internal and external factors, for important work in this area has recently been done by Edwards and Steins (1998), van Ginkel (1998) and Buck (1998). However, the examination of extractive reserves in the context of developments in Brazil and the international arena, which I undertake in the next five chapters, will provide a more detailed insight into this issue.

A preliminary framework for examining the joint use of natural resources was depicted in Fig. 1.1. A more sophisticated version of this framework is presented below, in Fig. 1.2. The conservation of a resource used in common by several individuals or families depends on the robustness of the common property regime or on the potential for such a regime to develop. The robustness of a regime depends on the jointness

Figure 1.2 Common property institutions and the conservation of CPRs (II)

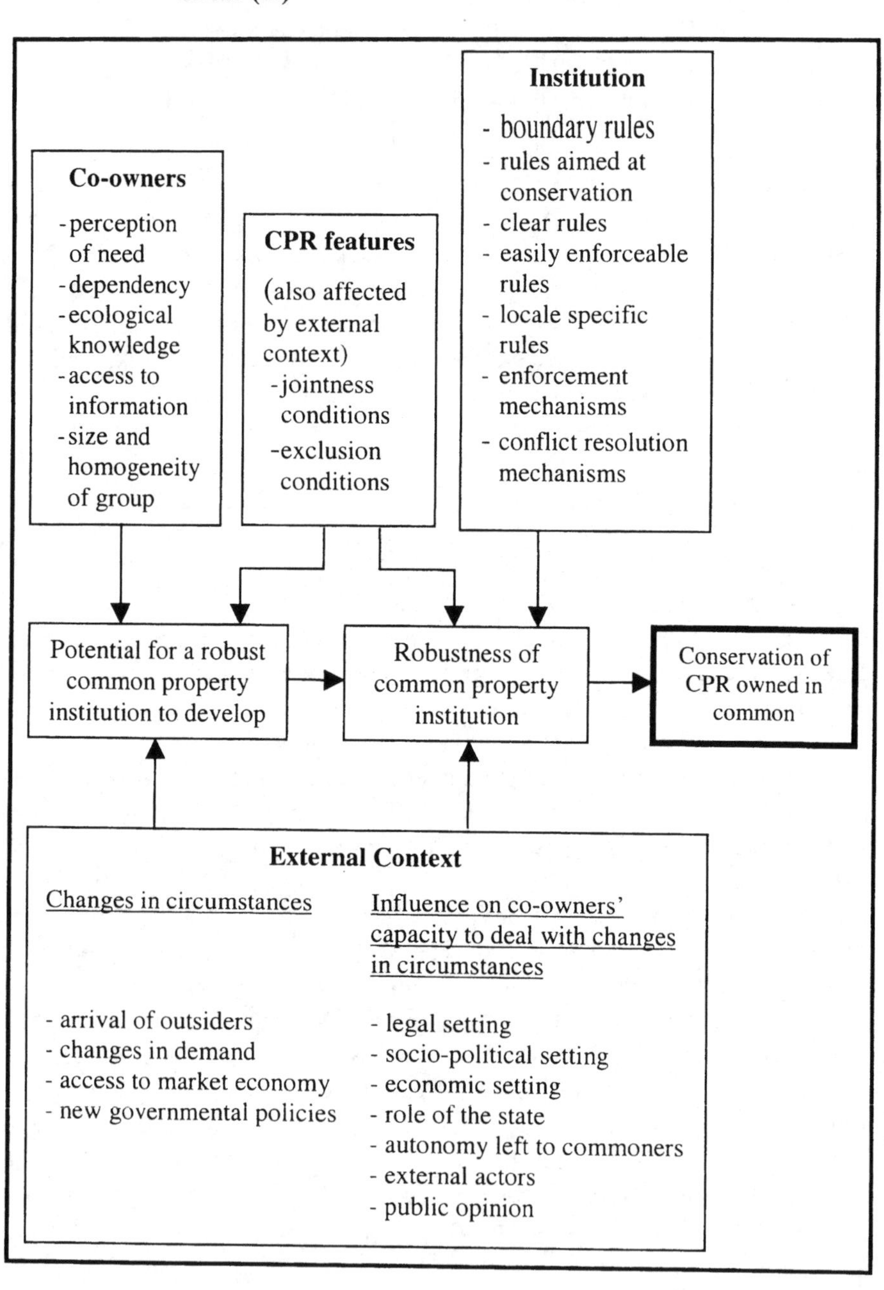

and exclusion conditions of the CPR and on the features of the regime: a robust regime must have clearly defined boundaries, rules aimed at harmonising the co-owners' use of the commons, and mechanisms for securing that these rules are followed. Whether a robust regime can develop if the need for it arises depends on the jointness and exclusion conditions of the common-pool, on the users perceiving the need, on their dependency on the natural resource system, on their knowledge of the resource, on their access to information, and on the size and homogeneity of the group. The external context influences the jointness and exclusion conditions of common-pools, the robustness of common property regimes and the potential for such regimes to develop.

Reviewing the external factors mentioned in the literature on common property, which are summarised in Figure 1.2, provided an indication as to what aspects of the external context tend to be important for individuals sharing natural resources. To carry out the research on extractive reserves, however, additional steps were taken to include the external context in the analysis. Based on the literature on extractive reserves, the main external factors that influenced them were identified. This follows a suggestion by Edwards and Steins (1998, p.8) who write that 'having identified a change in outcome to the CPR, the researcher can attempt to trace it back to a local or remote contextual factor by focusing on the choice sets available to individual users of the resource'. Yet it is often unclear which remote external factor determined a particular change in outcome. Frequently, changes occur because of a combination of factors and the researcher, having identified one of them, may be unaware of other external factors equally important. To secure a more comprehensive, although by no means exhausting, analysis of external factors, I include a review of the evolution of national and international policies in relation to the Brazilian Amazonia from the late 19th century, which is when the rubber tappers arrived in Amazonia, until the present day. The external factors identified in the theoretical and case study literature served as a starting point for this review, which is the subject of the next chapter.

Notes

[1] A fourth solution is the development of artificial markets. This system, however, is generally advocated for global common-pools such as the atmosphere rather than for local resources and is thus not discussed here.

[2] In the Middle Ages, for example, the owner of a manor had the right to pass on the land to his children and the right to receive an income from the peasants living

in the land, but the latter had the right to cultivate 'their' land, a right that the lord had to respect. In 17th century England, landed property was also limited to certain uses only, and generally, it did not include the right to transfer the landed property rights either by selling them or by bequest.

[3] Scholars who acknowledge that common property as defined above can secure the sustainable use of a CPR, use nevertheless the expression 'common property resources' to refer to open access resources. The term 'common property resources' often serves both to define two different institutional arrangements, and to designate the class of resources, independently of the institutional arrangement under which they are held (Buck, 1988; Berkes and Farvar, 1989; Hooker, 1994).

[4] Common property is thus different from the case of several individuals owning something as an artificial person, since under common property there is a combination of common and individual rights. It is not the same as corporate property either, in which case the shareholders have rights over part of the resource only, the shares, whereas in the case of common property the owners have rights over the whole resource (Reeve, 1986).

[5] This example refers to sedentary rather than nomadic pastoralists. In the second case, rules for establishing the boundaries of the resource (the pasture) may be as difficult to enforce as those of a fishery.

[6] See Keohane and Ostrom (1995) for a more detailed examination of heterogeneity and common property.

[7] Regime theory is often considered the counterpart of common property theory at the global level (Keohane, 1995; Keohane and Ostrom, 1995; Vogler, 1995), but not because it examines local management of CPRs from a global perspective. The two bodies of theory share similar assumptions (e.g. institutions matter) and analyse dilemmas that are alike. Like joint users of a CPR, national governments must find means to harmonise their actions with those of other actors without the help of an external agency at a higher hierarchical level (Keohane and Ostrom, 1995). This means that theorists on both international regimes and common property regimes examine ways of addressing the free-rider problem and ensuring compliance with established rules. However, whereas common property theory examines local issues, international regime theory, as the name indicates, focuses exclusively on international matters and it does not provide any framework to examine resource management at the local level.

Given the similarities between the theory on common property institutions and international regime theory, a potential method of analysis could have been to use common property theory to examine local factors and regime theory to examine the external context. Arguably, a framework integrating the two bodies of theory could have been developed and used to examine the interaction between internal and external factors. This approach, however, presents an important problem, which is that the utility of regime theory to examine the developments in relation to Amazonia during the 1980s is limited. Regime theory focuses on developments in the international arena whereas, for the joint users of a CPR, developments within the national arena are also likely to be important. Moreover, regime theory tends to assume that states are monolithic entities, and it is not possible to explain the developments that occurred in Amazonia during the 1980s without considering the dynamics between a range of different actors within Brazil.

[8] The 'successful' commons that have lasted for long periods were those 'that had not experienced drastic changes in the local economy that made non-traditional and easily individualised uses of the commons even more efficient than collective uses' (McKean, 1992, p.254).

2 National and International Developments: Their Impact on Brazilian Amazonia and its People

Introduction

Natural resource management is at the core of the Amazon rural economy. The livelihoods of over two million people in the region are based on their natural resources, forests and waterways. This is a conservative estimate, which includes only those individuals who depend directly on fishing or extracting forest products, such as rubber, Brazil nuts, and tropical fruits. The remainder of the rural population, another two million people, depends primarily on agriculture, but derives as well some income from forest products and, given the specific characteristics of Amazonian soils which are discussed later in this chapter, small farmers also need to engage to some extent in natural resource management.

Many of these people have lived in virtual isolation for long periods because of the inaccessibility of much of the region; even now, outside the cities, many communities are only accessible by small boat or on foot. At regular intervals, however, externally provoked events have broken their isolation and brought radical changes in their lives. The arrival of the Portuguese in the 16th century, for example, apart from widespread destruction of native cultures also brought with it international demand for the dyewood Pau Brazil, a native Amazonian produce whose commerce transformed the local and regional economy. Pau brazil was virtually depleted and until the rubber boom local populations in the region were, once again, left alone. In the late 1800s, the discovery of vulcanisation – a process that makes rubber's elastic properties permanent – and the growth of industrialisation generated a boom in the world demand for rubber. Amazonia had the virtual monopoly supply of rubber and the socio-economic structure of the region became centred on the rubber trade. Part of the local population was drawn into the rubber trade but, given the shortage of labour in the region, rubber traders promoted the migration of people from other regions to work as tappers on large rubber estates. They constituted the rubber tapper population, the central subject of this study. In the 1920s, rubber plantations developed in Southeast Asia and the Amazon rubber trade declined. Local populations were then again left to their own devices, to manage their resources with little or no interference from the government or other external agents until, in the 1960s, the government initiated a set of policies aimed at the 'development' of the region.

This chapter reviews developments in Amazonia from the time of these policies onwards, in order to understand the context in which resource users have had to take decisions. Chapter 1 postulated that the external context influences common property regimes because it can trigger changes in the circumstances to which co-owners must adapt, and because it can hinder or enhance their capacity to deal with such changes. During the period examined in this chapter (1960s-1990s), there were important alterations of the regional, national and international settings, which triggered radical changes in the circumstances of rubber tappers and other forest dwellers, and which influenced their capacity to deal with such changes. In examining the regional, national and international context of resource users, particular attention will be paid to the external factors identified in the literature: construction of roads, changes in government policies, changes in the socio-political and economic context, and alterations in the legal setting. Yet rather than analysing each of these factors individually, the review of the external context is presented in a chronological form. The development of a common property regime is a dynamic process, which takes place over time rather than at a particular point in time. Presenting the external context in a historical form provides more scope for assessing external factors at all stages of the process of development of the regime, and makes it possible to discern relevant external factors that may not have been identified in the theoretical literature.

The chapter begins by explaining the traditional property rights structure of Amazonia. It then turns to examine the government policies of the 1960s and 1970s, showing how they disrupted the regional property rights structure, and, in so doing, trigged high levels of deforestation. Subsequent sections examine the international setting: how in the 1980s interest in the conservation of Amazonia developed, and what was the impact of this interest at the national level.

Map 2.1 Legal Amazonia

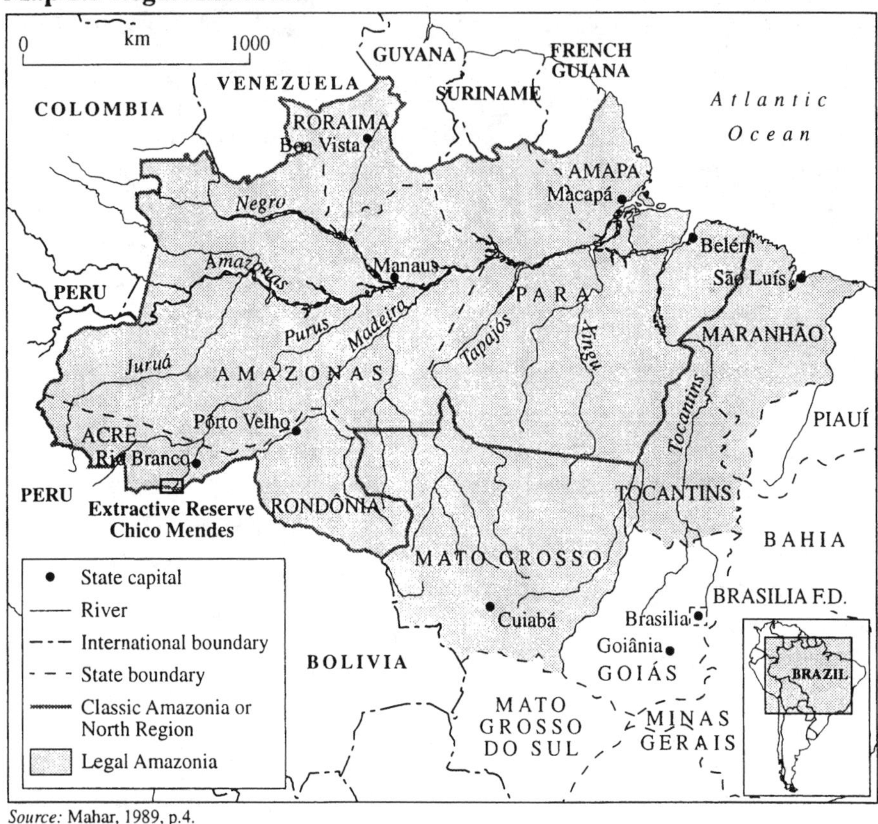

Source: Mahar, 1989, p.4.

The regional and national settings

Landed property rights in Amazonia

In many ways, landed property rights in Amazonia have always been different from landed property rights in the rest of Brazil, and in most of the western world. The traditional system of landed rights, that is, the system in place when the government began the occupation of the region in the 1960s, performed, nevertheless, the function of property rights systems in general. What belonged to who was specified, and people's property rights were respected. This may have been because resources were abundant, because society actively backed up the property rights structure of the region, or because of both. Whatever the reason, the fact is that people had secure access to their resources and structured their actions accordingly. Government policies disrupted this system, creating a situation of chaos in which traditional and modern concepts of property had to coexist with little or no arbitration from the state. The traditional property rights of Amazonia, however, were not only the result of an endogenous process, but of a combination of internal and external factors.

Two features characterised the traditional property rights system of Amazonia. The value of land was based on the resources of the land rather than on the land surface; this perception of rights cut across different groups, including peasants, rubber tappers and large private landowners, such as the rubber barons, whose estates were defined according to the number and quality of the rubber trees rather than the area they occupied (Cardoso and Muller, 1977; Schmink and Wood, 1992; Sawyer, 1984; Santos, 1984; Branford and Glock, 1985). One reason why land was valued according to its resources is that the Amazon economy has always been centred on the extraction of products from the forest (e.g. pau brazil, brazil nuts, rubber). As land in the region is abundant but commercial plant species tend to concentrate in relatively small areas, it was possible to own a large area of land with little or no value. The scarce resource was therefore what was on the land rather than the land itself, and consequently that was what determined the value of landholdings.

The other defining characteristic of the traditional property rights structure of Amazonia was the existence of a high proportion of holdings without legal property titles (Martins, 1980; Schmink and Wood, 1992). Although landholdings were not legally registered, they were recognised by society, and all relevant actors, the owners themselves and their neighbours, knew the limits of the landholdings. Many rubber barons did not have legal titles to their lands and the delimitation of their states was given by rivers and neighbouring estates (Branford and Glock, 1985; Duarte, 1986; Basilio, 1992). Peasants and extractivists generally lived in 'public lands' (*terras devolutas*), or on rubber estates that had been abandoned by its owners when the rubber trade collapsed. In the Amazon, much land was public but not managed by the state; people living on public land had no legal tittles to it, but they were not living there illegally. The dwellers of abandoned rubber estates were in a similar situation: they did not have any legal titles, but had some rights to the land where they lived and worked. Another ambiguous arrangement was the *aforamento*, a perpetual lease to exploit forest resources that fell into 'a nebulous realm between private and public property' (Schmink and Wood, 1992, p.64). The abundance of land in the region – due in part to its isolation from the rest of the country because of lack of transport connections – and the fact that land in Amazonia was hardly integrated in the national land markets, can essentially explain the absence of titled property rights.

Between the end of the rubber boom in the 1920s and the 1960s, the government was largely indifferent with regard to the use of natural resources in the Amazon region (Nuggent, 1993; Schmink and Wood, 1992; Allegretti, 1989). Local populations developed institutions based on the physical and social characteristics of the region, rather than on development patterns from temperate regions or international demand for certain products (Nuggent, 1993). These institutions had hardly any legal support and their existence was ignored by the federal state. Peasants and extractivists were in general untitled occupiers of their plots *(posseiros)*. According to Brazilian Law (*Estatuto da Terra*, 1964, law 4504) the person or family who has lived on a plot of land for over a year and a day, using the resource productively has usufruct rights to the land. The size of the plot must be of at least 0.5 sq. km depending on the number of family members able to work on the plot and on the quantity of land they need to make a living (Martins, 1991; Basilio, 1992; Schwartzman, 1992). *Posseiros*' rights (informal leaseholder's rights), however, applied only to individual or family ownership of land; although many traditional populations in the region had land tenure systems which involved also common property rights, the latter were ignored by Brazilian law (Ianni, 1979; CDEA, 1992). Until the early 1970s, however, it can be assumed that lack of legal and state support was not a problem for the functioning of these regimes. Given the isolation and sheer size of the region, commoners' rights were seldom threatened by non-owners attempting to take over their resources and, consequently the protection of common-pool boundaries was not essential.

Change in the national setting: the government policies of the 1960s, 1970s and 1980s

From 1964 onwards the socio-political and economic context of the Amazon population began to change. The Brazilian government developed a new approach to the region whose aim it was to integrate the Amazon in the national economy. Initially, the state was still indifferent with regard to common property regimes, in the sense that it never attempted to take control over jointly owned resources; the existence of common property regimes was simply ignored. The new government policies, however, included constructing highways across the region and connecting Amazonia with the rest of the country, promoting migration to the region, and encouraging investment in it. These policies led to a radical change in the external setting of resource-users in the region: they created a situation of open access with a high degree of chaotic competition for land and resources, which resulted in deforestation and resource degradation.

The government plans for developing Amazonia began with the construction of the new capital, Brasilia, closer to Amazonia than to the traditional centres of development in the coast, and with the opening of the Belém-Brasilia highway, which provided for the first time a terrestrial link between the region and the rest of the country. In 1964, the government set up its first national plan for the region: Operation Amazonia. To boost industrialisation and commerce in the region, it established the Manaus Free Zone, and created SUDAM (Superintendency for the Development of Amazonia), a public agency for the administration of fiscal incentives for investment in the region. It also engaged in the construction of a vast network of roads, airports, and other transport infrastructure, which by facilitating access to the region changed the exclusion conditions of the Amazonian forests. Roads and fiscal incentives attracted people to the region, altering the scarcity conditions of the forest.

In 1970, in the context of the National Integration Programme (PIN), the new government plan for the region, even more people were attracted to Amazonia; PIN's aim was to encourage migration into the region, especially along the newly opened TransAmazonian highway. Most migrants came from the Northeast of Brazil, where a very severe drought had increased tensions among landless peasants and *posseiros* on the one hand, and large landowners on the other. As Amazonia was sparsely populated and many of the existing property rights arrangements were not legally registered, the region was considered 'empty'. The colonisation of an empty zone would provide land for landless people – 'land without people for people without land' was the slogan of the government – and diminish tensions in the Northeast (Lisansky, 1990; Sawyer, 1990).

From 1974 onwards, the government switched to a different approach for the colonisation of Amazonia, and set up POLAMAZONIA, that focused on 15 development poles largely geared towards exports so as to pay for the external debt (Hall, 1987; Diegues, 1992) and to 'fuel Brazil's economic development' (Goodman and Hall, 1990, p.5). Capital investments in mining, large-scale farming and cattle ranching were encouraged through tax exemptions, subsidies and other facilities aimed at attracting investors to the region.

These three programmes altered the socio-economic context of the region's inhabitants in several ways. They attracted many people to Amazonia. This alone would have enhanced competition for resources, hence weakening existing common-pools' boundary rules, but two other factors made this competition particularly fierce. First, newcomers' use of land required the removal of the trees; joint use of resource by traditional users (whose livelihood depended on the standing trees) and newcomers was therefore impossible. Second, property rights were unclear and were unprotected: local people's rights were frequently ignored, and for newcomers, deforesting was a means of securing property rights. An intense struggle to obtain landed property rights ensued, and this resulted in high levels of deforestation. For Amazonian resource users, deforestation deprived them of their livelihood, but it also attracted international attention to the region and this contributed to further alter their political and socio-economic context.

How much deforestation occurred is a highly debatable issue, since the political agenda of organisations presenting the data influences the levels of deforestation claimed (Kolk, 1996). Besides, there are different definitions of forest and deforestation, different methodologies for estimating deforestation rates, and lack of accurate and up to date data. There is, however, virtual consensus that from the late 1970s onwards, and especially during the 1980s, much of the forest in Amazonia was being cleared[1].

The figures for deforestation presented here refer only to the complete destruction of the forest cover, when no trees are left standing. Some estimates suggest that between 1975 and 1980 the deforested area of Amazonia increased from 29,000 sq. km to 125,000 sq. km (Mahar, 1989), corresponding to 0.6% and 2.5% of the region. FAO (United Nations Food and Agricultural Organisation) indicates that during the 1980s, approximately 205,000 sq. km were cleared (FAO, 1994). In 1987 deforestation was particularly high, and estimates suggest that by 1988 between 8% (Mahar, 1989) and 12% (Myers, 1989) of Amazonia had been cleared. The percentage of deforestation in relation to the entire rainforest was still low by comparison with other tropical regions, but the rate of deforestation, was worrying because it was exponential rather than linear (Fearnside, 1985).

As will be seen in later chapters, the high deforestation figures of 1988 were one of the factors that triggered international concern with the Amazon forest and, indirectly, world wide interest in the rubber tappers' plight. In the late 1980s, there was a slowdown in deforestation, which was initially assumed to be related to the national and international reaction to the 1988 deforestation figures. From 1991 onwards, however, deforestation increased again and, overall, it has been higher in the 1990s than in the 1980s. Between 1978 and 1988, the average annual deforestation was of 17,000 sq. km, whereas between 1988 and 1997, it was of 20,400 sq. km (Fearnside, 1997; 1998). Deforestation was particularly high in 1995, when 29,000 sq. km of forest were cleared; since then, forest clearing has decreased again and estimates for 1998 suggest that in that year deforestation totalled 17,000 sq. km (Fearnside, 1997; 1998).

The distribution of deforestation among the different Amazonian states is also important to consider, since government initiatives were not evenly distributed and not all states were equally affected by deforestation. While the state of Amazonas, which is located in the centre of the region and thus of difficult access, remained essentially intact throughout the 1980s, 14% of Rondônia and 47% of Maranhão, which are frontier states, were cleared by 1990. Acre, the state where the rubber tappers' struggle began, also had a high level of forest destruction during the 1980s: its annual deforestation rate during that period was of 45%. By contrast, in 1991 (thus after the establishment of extractive reserves), the area of forest cleared in Acre corresponded to only 3% of the total area cleared in that year. Most deforestation in that year concentrated in Mato Grosso, where 26% of the total deforestation for 1991 took place (Fearnside, 1998).

One of the reasons why government policies resulted in such high levels of deforestation is that they encouraged activities that are ecologically unsuitable for the region. Activities that require clearing the forest cover, such as cattle ranching and agriculture, are unsustainable because the soils in the region are extremely poor[2]. Therefore, once a piece of land is cleared it cannot be used indefinitely for that activity – sooner or later the owner will have to move to new areas and clear more forest. Contrary to what happens in temperate regions – where nutrients are harboured in the soils – in tropical rainforests nutrients are concentrated in the vegetation[3]. When a small area of forest is cleared (1 or 2 hectares), the nutrients remain in the system: they go from the cleared area go into the next patch of forest and, later, small animals or the wind transport the seeds back to the clearing permitting the regeneration of the forest. When, however, a large area of forest is cleared, it cannot regenerate because the nutrients leave the ecosystem. They go into the crop or pasture and leave the system with the harvest or when the animals are sold. The majority is lost from the system because once the land is left bare, the capacity of the soil to absorb rain diminishes, and the increased runoff carries the remaining nutrients away. The impact of tropical rains, which are particularly strong and thus difficult for the soil to absorb, combined with the heat, and the existence of slopes in much of the region, creates then the problem of erosion[4].

Pasture creation was the main direct cause of deforestation (Hecht and Schwartzman, 1988), and the main source of problems for the rubber tappers. According to estimates by Browder (1988), by 1983, pasture formation was responsible for approximately 70% of the deforested area in Amazonia Legal; one third of it occurred in livestock projects subsidised by the government through SUDAM. Cattle ranching is, moreover, one of the worst land uses for Amazonia because it reduces the mineral stocks of the ecosystems, which leads to the development of weeds and to soil compactation, hence, the pasture carrying capacity diminishes fast and ranching has negative returns after approximately ten years (Goodland, 1980; Hecht, 1983, 1985; Hecht and Cockburn, 1989; Fearnside, 1990).

Mining also had considerable environmental impact. A good example is the Grand Carajás Programme, established in 1980 as part of POLAMAZONIA to exploit mineral reserves in the north of the region, covering an area of 800 – 900,000 sq. km or 10% of the territory of Brazil (MMA, 1995). This project involves the exploitation of a range of different minerals, such as iron, copper and cassiterite; a railway linking various iron melting plants to the sea, in São Luis de Maranhão; an agricultural project; and a dam, Tucuruí, to provide energy to the industries. Although the company running the Grand Carajás Programme, Company Vale do Rio Doce (CVRD), set up some environmental programmes to mitigate the overall impact of the mining complex, the environmental destruction in the surrounding area has been considerable. The construction of the dam involved flooding 2,435 sq. km of forest (Diegues, 1992); surrounding forests were cleared for the use of the melting plants (Rich, 1994); in addition, the Programme has served as a catalyst for attracting migrants to the region, resulting in levels of deforestation in the area of southern Pará and an upsurge of rural violence in conflicts over land (Hall, 1989).

Migration to Amazonia was the second most important cause of deforestation in the region. Taking together the colonisation projects along the TransAmazonia and the Polonoroeste, a colonisation programme in Rondônia within the context of the POLAMAZONIA, by 1983 migration by small peasants was responsible for 11% of total deforestation in the region. This figure represents only part of deforestation by migrants, since the latter arrived not only in colonisation projects but also in areas where roads had been opened, and where mining and logging opportunities existed (Fearnside, 1990). Contrary to traditional populations in the area, migrants, after clearing forest for cultivation, fail to leave the land fallow during a sufficiently long period that could allow the forest to regenerate; they plant shortly afterwards (obtaining lower yields) or transform the land into pasture (Fearnside, 1990). Colonisation along the TransAmazon highway is a case in point of the problems colonists encountered in Amazonia. The aim of the TransAmazonia colonisation plan was to settle 100,000 families by 1976, but by 1978 less than 10% of the families had been settled as originally planned. The colonists lacked the necessary logistic support from INCRA, the government agency responsible for the colonisation plan, and were unable to deal with the characteristics of the tropical environment. Apart from the incidence of tropical pests and diseases, the cleared land did not give the expected results because of the poverty of the Amazonian soils (Moran, 1981; 1990; Browder, 1988).

Deforestation occurred mainly along the highways that were opened in the region during this period (1970s and 1980s) such as the Belém-Brazilia highway, the TransAmazonia, and, the Cuiabá –Porto Velho highway (Mahar, 1989; Myers, 1989). For traditional communities in the area, the opening of the highways represented important changes in their circumstances, as will be shown in Chapter 3 when discussing the impacts of the Cuiabá-Porto Velho highway (BR-364) on the rubber tappers living nearby.

Encouraging agriculture, mining and livestock production affected local population in three interrelated ways. First, these uses are incompatible with the conservation of the forest because they require its removal. Second, given the characteristics of the Amazon soil, the deforestation produced by the newcomers impacted not only the deforested area itself but also its surroundings. Whatever use of their resources individuals living in the radius of the deforested area make, and whichever property rights regime they have, if they clear a large track of forest they will threaten the sustainability of their own resources and of those belonging to their neighbours. Third, the new comers competed for land between themselves and with the traditional inhabitants; in a context of insecure property rights, such competition was in itself a major cause of deforestation.

The approach of the government to the region led to a dramatic increase in land demand (see Figure 2.1) and as property rights were precarious, individuals (newcomers in particular) cleared the land to obtain legal titles to their holdings. Whereas before the 1970s, land in the Amazon had been abundant and thus informal property rights institutions could exist with little or no state and legal support, this had become impossible. The government policies triggered a change in the exclusion conditions of common-pool resources in the Amazon: to prevent the depletion of their resources by outsiders, commoners would have to strengthen their boundary rules.

As can be observed in Figure 2.1, several factors led directly or indirectly to an increase in land demand in Amazonia. 'Push' factors operating in other parts of Brazil thus also encouraged migration to Amazonia. Apart from the regular droughts and land concentration problems of the North East, the development of agribusiness and export orientated wheat and soybean production in the South of Brazil also served as incentives for landless peasants to become colonists in Amazonia (Goodman and Hall, 1990). Private investors were attracted to the region by the government incentives to buy land in Amazonia, and because of the newly opened roads. In Brazil, at the time, inflation was extremely high and this made buying land a valuable investment; given the government subsidies for buying land in the region, land tended to appreciate in market value at a higher rate than inflation, becoming thus a secure edge and the best protection against inflation (Fearnside, 1989b; Browder, 1988). High demand for land, fiscal incentives for buying it, and inflation, led to land speculation, which in turn further increased demand for land in Amazonia. Finally, a rise in demand naturally pushed up the price, further increasing land demand.

The increase in land demand led to deforestation because, first, the market price of cleared land was higher than of land covered by forest; many new owners wanted to use the area for large-scale agriculture or pasture, and they preferred to buy land already cleared because otherwise they would have to clear it themselves. Second, under Brazilian legislation at the time, clearing the forest constituted a productive investment of the land and was often necessary to obtain a legal title, which in turn was essential to gain access to government incentives, such as subsidised rural credit, as well as to commercialise land (Schwartzman, 1992).

Figure 2.1 Demand for land in Amazonia

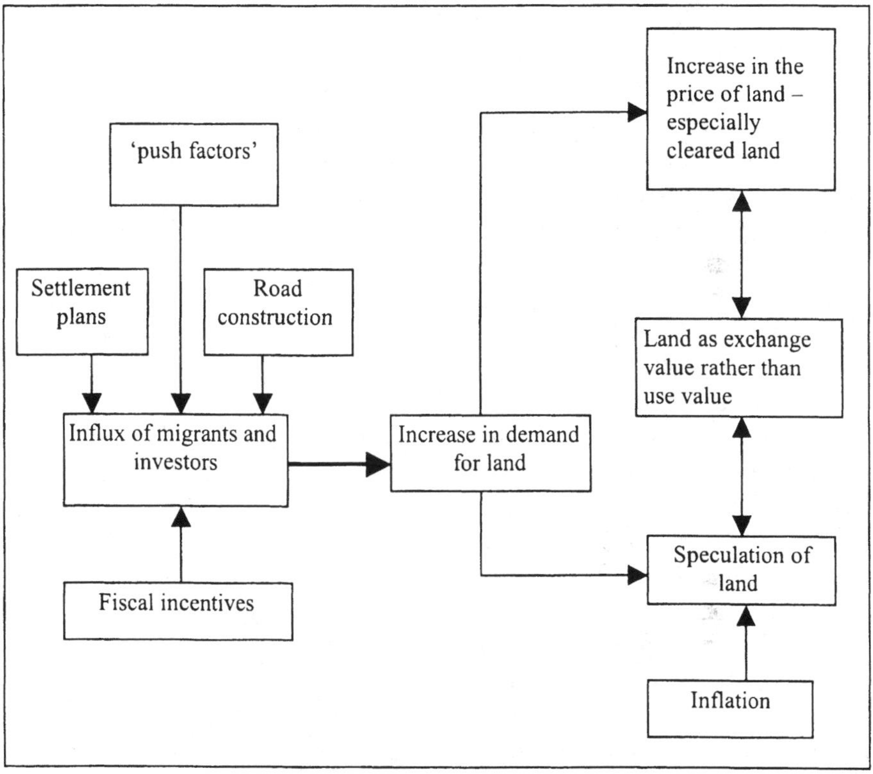

'push factors'

Settlement plans

Road construction

Influx of migrants and investors

Fiscal incentives

Increase in demand for land

Increase in the price of land – especially cleared land

Land as exchange value rather than use value

Speculation of land

Inflation

Sources: Fearnside, 1989b; Gross, 1990; Hecht and Cockburn, 1989; Branford and Glock, 1985.

Third, property rights, especially of both traditional and newly arrived *posseiros*, were unprotected, and this led to a violent conflict over land, which catalysed further deforestation by newly arrived peasants. Large landowners wanted *posseiros*, peasants, extractivists, and Indian populations out of their estates because their presence diminished the market value of the land, and created substantial problems for obtaining land titles and subsidies.

Many *posseiros* were expelled through legalistic means, based on their ignorance of their usufruct rights to the lands they occupied, or their lack of formal documents proving their leaseholders' rights. Even if aware of their rights, small farmers were frequently unable to follow the necessary bureaucratic procedures for securing legal ownership of the land; many of them were illiterate and they could ill afford to travel to the nearby city to register their plots. When unable to use legalistic means ranchers and speculators resorted to violence (Brock and Hesler, 1993; Martins, 1991; Americans Watch, 1991; Branford and Glock, 1985). The eviction of small peasants further increased deforestation. *Posseiros* knew there was a considerably risk that they may be evicted from their plots, and that the only compensation they could obtain for it would be based on the improvements they had made to their land. As clearing the forest cover was legally an improvement, they cleared as much forest as they could. As colonists moved further into the forest, they cleared more land in a never-ending cycle; some families moved from plot to plot four to five times (Branford and Glock, 1985).

In sum, deforestation in Amazonia can be explained as the result of an intense struggle to obtain and secure landed property rights. This struggle was a 'tragedy of the commons', in the sense that clearing the forest was often the best strategy for individual actors, but not for society as a whole – society in this case being humankind. Before turning to examine the global importance of the rainforest, however, it should be noted that also within Amazonia clearing the forest was not the best individual strategy for all actors. Whereas some individuals benefited from deforestation, such as cattle ranchers, others only suffered from it, especially traditional populations who depended on forest resources for their survival; the destruction of these communities' common-pool resources occurred because their boundary rules were insufficiently strong to exclude outsiders from occupying and destroying their resources, and not because they overused their common resource. The 'tragedy' in Amazonia was thus not so much an issue of individual strategies versus the common good, but rather of different groups having different interests.

The international setting

Chapter I postulated that in an interdependent world external factors have an impact on common property regimes defined as geographically bounded wholes (van Ginkel, 1998). For regimes located in the Amazon rainforest, this argument appears particularly relevant for two reasons. First, the boundaries of a local common-pool can never fully match the ecological boundaries of the resource because, from a strictly ecological point of view, the resource is too large. Given the interdependency of the ecological system that constitutes the rainforest, the conservation of local common-pool resources is systematically influenced by the use of the forest in neighbouring areas. Second, it can be expected that actors from the regional and global populations will have an interest in how people in the region manage their resources because they are also affected by it: they benefit if resource users conserve the forest and suffer the consequences of biodiversity loss and climate change if the forest is destroyed. Yet, the global ecological importance of Amazonia was not the only reason why the international community took an interest in the region or why the international context should be included in the analysis of local resource management.

Large-scale deforestation can influence the global climate because it contributes to the global warming of the Earth. Increased concentration of gases such as carbon dioxide (CO_2), methane, CFCs, and nitrogen oxide in the atmosphere raises the average temperatures of the Earth, the so-called 'greenhouse effect'; this can trigger sudden changes in temperatures around the world leading to droughts in some regions and floods in others (Brack and Grubb, 1996; Paterson, 1992). Given that forests absorb and retain carbon dioxide, the gas that contributes most to the greenhouse effect[5], when forests are destroyed, they naturally stop acting as carbon sinks and, moreover, they release the carbon stored in the plants back into the atmosphere. According to World Bank (1991c, p.36) estimates, the contribution of deforestation in Brazilian Amazon to the greenhouse effect is approximately 4% of total CO_2 emissions; studies by Fearnside (1997) indicate that the deforestation that occurred in Legal Amazonia in 1990 represents 5% of the total global emissions from both deforestation and fossil fuel sources in that year[6].

Another way in which large-scale deforestation affects the global climate is through its disruption of the hydrological cycle of the region. This can degrade the remaining forest and, given the large amount of water in Amazonia (the Amazon River alone carries 20% of the world's freshwater flow), and affect both the regional and global climate[7]. Some models suggest that alterations of the hydrological cycle may result in droughts in the Central Plateau and Southern regions of Brazil, and a decrease in rainfall in the temperate parts of the world[8] (Salati, 1985; Fearnside, 1990; 1998; Lovejoy and Salati, 1983).

Amazonia is also globally important because it harbours one of the highest biodiversity reserves of the world. Biodiversity is the variety of all living organisms on Earth, including the ecosystems where they live, 'variability among living organisms from all sources, including, *inter alia*, terrestrial, marine and other aquatic ecosystems and the ecological complexes of which they are part; this includes diversity within species, between species and ecosystems' (article 2, definition of biological diversity, Biodiversity Convention, in Johnson, 1993). Estimates of Amazonian biodiversity suggest that there are between 1 and 15 million species in the region – estimates for the biodiversity of the entire planet vary between 5 and 30 million (Browder, 1989; CDEA, 1992)[9]. Biodiversity is of paramount importance for humankind because it is 'the common immunisation system of global life and of life connected industries' (Liepietz, 1995, p.121). The diversity of species, and the genetic information that exists in each living organism allow scientists to discover, for example, new antidotes to pests and diseases, or to develop new crops. This rich variety of species is easily destroyed: most species are endemic to particular areas and cannot live in isolation – the destruction of a relatively small patch of forest can thus result in considerable destruction[10]. It is estimated that only a tenth of the different living organisms there are on Earth are known, and of the already classified species, only a small part of their information has been documented. Deforestation – and the accompanying destruction of species – represents therefore the loss of valuable information that cannot be later retrieved.

The Amazon region, however, has been always important in ecological terms, but only in the 1980s public opinion in the North took an interest in it. In the 1970s, international actors' role with regard to Amazonia was undoubtedly less important than a decade later, but it existed nevertheless and was totally unrelated to the value of the standing forest.

International actors in Brazilian Amazonia – 1970s

During the 1970s, the international actors' role in the region was related to the importance of the region in economic terms in the strict sense. Broadly speaking, during this time they never voiced any concern with the conservation of Amazonia, but rather supported the government approach to the development of Amazonia. To what extent their support shaped events in the region is, however, debatable: it is 'a moot point whether [pressure from overseas interests] has actually undermined Brazilian development choices in Amazonia, or simply fitted into an existing set of national priorities' (Hall, 1989, p.253-254).

On the one hand, it has been argued that international capital was determinant in the environmental degradation of the rainforest (Cota, 1984; Kovarick, 1995). The Grand Carajás, for example, was partly financed by loans from the World Bank, the InterAmerican Development Bank and foreign private investments, especially from Japan (Sauthchuck et al, 1980; Pinto, 1982; Kovarick, 1995). The Jari project is another case in point: covering an area of 12,000 sq. km, it was set up by a North American magnate, Ludwig, who planned to develop plantation forestry, pulp processing, mining, livestock and irrigated rice cultivation in the middle of the rainforest, in an area partly inhabited by rubber tappers.

On the other hand, there are strong arguments supporting the view that, in the 1970s and 1980s, international actors and capital contributed to the process of destruction of the forest, but did not determine it. Hecht and Cockburn (1989) point out that most land clearing in the region has been for pasture creation, an activity carried out mainly by Brazilian capital. Foreign demand for beef did not play a role either, since Amazonian beef is only consumed domestically (Hecht, 1983). Nor was external debt the triggering factor in the deforestation process: Brazil's exports amount to less than 10% of its GDP and almost 50% of these exports come from textiles and manufacturing activities; agricultural exports come in their majority from outside Amazonia (Hecht and Cockburn, 1989). The Jari project was indeed set up by foreign capital but when in 1982 a controlling interest in the estate was sold to a consortium of Brazilian firms ecological destruction continued (Fearnside, 1990; Hecht and Cockburn, 1989).

Hurrell (1992) and Kolk (1996), who have examined deforestation in Amazonia in the context of international relations, also argue that there is not a direct link between international factors and deforestation in the region. The state took the lead in the development of Amazonia and although it was supported by international capital, the latter had a clearly subordinate position. However, they consider that 'through the incorporation of the Amazon in the Brazilian political economy [...] the international linkages and the articulation of foreign influence can be understood' (Kolk, 1996, p. 73). International loans contributed to the debt crisis – like Grand Carajás, many of the projects of POLAMAZONIA were partly financed by loans from the World Bank and the InterAmerican

Bank, and the Banks' support was an incentive to attract foreign private investments. The debt crisis in turn stimulated inflation, which as mentioned earlier has been one of the incentives for clearing land (Hagemann, 1994).

While in the 1970s, international actors' role in Amazonia worked in the same direction of that of the government, supporting their policies and contributing to the destruction of the rainforest, in the 1980s there was a radical change. International capital continued investing in the region, but as international concern with deforestation in the region developed, a series of conflicts between Brazil and the international community ensued. The global ecological importance of Amazonia, however, was not alone responsible for this shift in the general attitude of international actors. Knowledge, information, economic differences, and political manoeuvring also helped to make Amazonia a global environmental issue.

Deforestation in Amazonia becomes an international political issue – 1980s

Concern with deforestation in Amazonia was part of a general trend during the 1980s, in which the focus of environmental interests gradually shifted from local to global issues. The coverage of several environmental disasters during this period, such as the nuclear accident at Three Mile Island in 1979, Chernobyl in 1986, desertification and accompanying famine in Ethiopia, as well as, by the end of the 1980s, burning of the Amazon rainforests, highlighted the importance of environmental matters outside the restricted circles of scientists and other specialists (Thomas, 1992). Growing scientific knowledge on global issues such as the greenhouse effect and the importance of biodiversity also played a role in highlighting the ecological interdependency of the Earth, and in particular the importance of conserving forests for the benefit of humankind. Finally, the actions of NGOs in publicising and campaigning for environmental issues firmly place environmental issues amid the concerns of the public and, consequently, in the political agenda of many countries (Bramble and Porter, 1992; Hurrell, 1992; Thomas, 1992; Rowlands, 1992; Prince and Finger, 1994; Kolk, 1996).

It was also during the 1980s that the relationship between environmental conservation and economic development began to be understood as interdependent rather than mutually exclusive. This, as will be seen in Chapter 3, had substantial implications for traditional resource users in Amazonia: traditional populations gained political advantage because their economic activities were environmentally sustainable. The link between environment and development had already been formally acknowledged at an inter-governmental level in 1972, at the United Nations Conference on the Human Environment in Stockholm. This conference was a milestone in the process of the recognition of environmental problems at the international level and lead, for example, to the creation of the United Nations Environmental Programme (UNEP). It took, however, another 15 years and the publication of 'Our Common Future' by the World Commission on Environment and Development (the Brundtland Commission), for the interdependency between environment and development to become mainstream. This interdependency was encapsulated in the concept of sustainable development.

Although the concept of sustainable development did not originate in the Brundtland report, it was through this report that it gained worldwide political leverage[11]. The Commission defined it as development that 'seeks to meet the needs and aspirations of the present without compromising the ability to meet those of the future. Far from requiring the cessation of economic growth it recognises that the problems of poverty and underdevelopment cannot be solved unless we have a new era of growth in which developing countries play a large role and reap large benefits' (WCED, 1987, p.40). To translate the concept of sustainable development into practical policy is problematic and, besides, what exactly the definition of the Brundtland Commission means has been highly debatable[12]. Yet in spite of the different interpretations of the concept (or perhaps because of the concept's potential for different interpretations) the term created a consensus about the link between environment and development. Environmental concerns are not necessarily obstacles to development, and development could and should occur in harmony with environmental conservation. In this context, the use of natural resources by traditional forest populations became paradigmatic of sustainable development; these communities use the forest to meet economic objectives but do so in a sustainable manner, that is, their activities do not destroy the resource. Moreover, in conserving their own resources, traditional forest communities in Amazonia also mitigate a globally important problem, the destruction of the rainforest.

The conservation of tropical forests was perhaps one of the most salient global environmental issues of the decade. Deforestation had increased dramatically during the 1970s and knowledge of the global impacts of deforestation was spreading fast. Two important international initiatives were taken to address the problem, the Tropical Forest Action Plan (TFAP) and the International Tropical Timber Agreement (ITTA), but their successes was limited. In part, this was because both initiatives largely ignored local resource users. Adopted at the World Forestry Congress in 1985, the TFAP was a global forest conservation and development program co-funded by the United Nations Food and Agricultural Organisation (FAO), the World Resources Institute (WRI), a Washington based NGO, the World Bank and the United Nations Development Programme (UNDP). Its objective was to increase the flow of forest aid to US$8 billion over five years, halt the destruction of tropical forests and facilitate sustainable use of forests (Thomas, 1992). The programme, however, failed to meet the planned objectives: during the 1980s deforestation rates increased in most of the tropical world, and in Amazonia, as shown earlier, they reached particularly high levels.

One reason for the lack of success of the TPFA in halting deforestation was that its focus was on forestry, while the causes of deforestation often lay outside the sector (Humphreys, 1996). In Amazonia, for example, lack of secure property rights, incentives to invest in cattle ranching and push-factors outside the region were more important promoters of deforestation than logging, which became a problem only in the 1990s. Another shortcoming of the TFAP was that there was a lack of harmonisation between national and international interests, and the projects were, according to many critics, donor-driven rather than geared towards the priorities of the local populations (Humphreys, 1996). The World Rainforest Movement (WRM) and the World Resources Institute explicitly criticised the programme because of its lack of harmonisation with local interests (Humphreys, 1996). WRM pointed out that the independent review commissioned by FAO treated tropical countries as:

> [M]onolithic entities, thereby obscuring critical conflicts of interest within these countries which are a major cause of forest loss... The Review conflates the interests of the inhabitants and governments of tropical countries ... (yet) it is the marginalization of the rural poor by the development process, and the expropriation of their lands and resources, which is one of the main engines of forest destruction (Colchester and Lohmann, 1990, p. 99).

This latter criticism is particularly relevant to the case of Amazonia, where, as seen earlier, a considerable extent of deforestation occurred as a result of government development plans for the region, which largely ignored the needs of both migrants and traditional populations in the region.

The ITTA is an agreement on the trade of timber, administered by the International Tropical Timber Organisation (ITTO), an international organisation formed by producer and consumer countries. In the late 1980s, Brazil did not have an important share of the world trade in timber, but nevertheless it participated in the agreement. Although the ITTO was originally planned to be only a commodity agreement, it included a sustainable development component, and in 1991, the signatories agreed that from the year 2000 onwards they would only trade in timber from sustainably managed forests (Kolk, 1996). ITTO has been criticised by Friends of the Earth who argue that it promotes trade at the expense of conservation, but, on the other hand, it should be noted that by promoting trade from sustainably managed forests, ITTO contributes to diminish predatory harvesting of forests (Thomas, 1992). Notwithstanding the controversy over logging, critics of the programme consider that project appraisal should not be done by the submitting countries but by ITTO itself, and that there is lack of consideration of the environmental and social impacts of the projects (Kolk, 1996; Thomas, 1992).

An international initiative that, contrary to the other two, had some success in halting deforestation was an NGO campaign against the project lending policy of the Multilateral Development Banks (MDBs). Aimed at showing the negative environmental consequences of the existing model of development, and centred on the connections between international and local environmental problems, the MDB campaign played an important role in stopping environmentally destructive projects, in particular in Brazilian Amazonia. Although not aimed exclusively at forest conservation, it brought the issue of deforestation into the attention of public opinion, especially in the industrialised countries, and catalysed interest in the plight of local populations. This in turn had important implications for the rubber tappers: it was in the context of the MDB campaign that they developed alliances with international actors which triggered their success in obtaining legal support for their property rights institutions.

The MDB campaign began in 1983 at the initiative of three US-based NGOs, the National Wildlife Federation, the Natural Resources Defence

Council, and the Environmental Policy Institute (the predecessor of Friends of the Earth), who were looking for 'ways to deal with the international dimension of the environmental problem which would preferably be understandable for a wider public' (Kolk, 1996, p.248). Their strategy for doing this, which turned up to be very successful, was to emphasise the impact that projects financed by the MDBs were having on local people and their environment. In cooperation with Southern and European NGOs, activists selected specific projects in developing countries, identified their social and environmental impacts and then, together with local communities' representatives, lobbied the MDBs to alter or cancel the projects; since much of the financing for some MDBs comes from the USA, they also lobbied the American Congress to alter the MDBs' policies. In Brazilian Amazonia, the NGOs focused on two large-scale projects: Grand Carajás and Polonoroeste.

The campaign against the Polonoroeste had significant implications for the rubber tappers. One of POLAMAZONIA's 15 projects, Polonoroeste involved paving the BR-364 highway from Cuiabá to Porto Velho (Rondônia), and setting up colonisation projects for small farmers who had been expelled from the South of Brazil due to land concentration (Hall, 1993). Partly financed by the World Bank, which between 1981 and 1983 granted loans of US $443 million, this project had disastrous social and environmental consequences (Rich, 1994). The paving of the highway attracted a much larger number of migrants into the region than expected and approximately half a million colonists arrived between 1981 and 1985. The infrastructure of the government colonisation agency, however, could not cope with such a high number of colonists and the latter did not receive the necessary logistical and financial help (Martine, 1990). Without support from the government and unaware of the poverty of the Amazon soils, colonists cleared large areas of forest for cultivation or speculation; conflicts over land became frequent and the deforested area in the region rose from 1.7% in 1978 to 16.1% in 1991 (Rich, 1994). In 1984, NGOs from Brazil, the USA and Europe sent a letter to the president of the World Bank setting out the serious consequences that the project was having, but the Bank ignored their criticisms. NGOs then decided to lobby the American Congress with the help rubber tappers' supporters: Mary Allegretti, an anthropologist who supported the tappers during their struggle for the establishment of extractive reserves, and Jose Lutzenberger, who later became the director of a government environmental agency that backed up the tappers' extractive reserves. Both Allegretti and Lutzenberger travelled to Washington to present evidence on the negative environmental and social impacts of the projects.

In 1985 the World Bank stopped disbursements for the Polonoroeste unless and until some social and environmental precautions were taken (Revkin, 1990; Kolk, 1996). Soon afterwards, however, the InterAmerican Development Bank (IDB) conceded a loan for the same project, now for paving the extension of the BR-364 highway from Rondônia to Acre. There were hardly any measures in place to prevent the environmental and social problems that had already occurred in Rondônia and a similar level of destruction could be expected in Acre. Environmental NGOs began lobbying the relevant bodies once again. By now, NGOs had made contact with the tappers and were supporting their struggle for secure landed property rights. In 1987, one of the tappers' leaders, Chico Mendes, gave evidence at the IDB and at the American Congress of the impacts that Polonoroeste had already had in Rondônia, and of the likelihood that the same could happen in Acre if preventive measures were not taken. In addition, Mendes asked for support and recognition of the tappers' common property regimes in the form set out in the Extractive Reserves proposal. Eventually, the IDB loan was suspended, subject to an agreement regarding the rights of Indians and rubber tappers in the area. As a response to this demand, the Brazilian government, since 1985 under a civilian president, José Sarney, had to ensure that the PMACI, a programme for the protection of the environment and of indigenous rights that had been requested by the IDB as a condition for providing the loan, was followed.

In 1988, international criticism of the Brazilian government policy in the Amazon reached a peak. The MDB campaign was largely responsible for this, but they were also several coincidental events that played a role. One was the publication of a report by the Brazilian Space Agency (INPE) showing that 80,000 sq. km had been burned in the previous year (Kolk, 1996). Taking the above figure as a starting point, and using exponential rates to estimate future levels of deforestation, some calculations suggested that if these rates continued, most of the region would be cleared by the year 2000. More recent calculations suggest that deforestation in 1987 may have been approximately half the above figure. Besides, it was also shown later that the level of deforestation in 1987 could not be used as an indication of deforestation in the future because forest clearing in that year had been the result of a specific combination of factors. The weather had been particularly dry in 1987, and expected changes in landed property rights had encouraged resource users to deforest more (1987 was the last year in which clearing land gave access to tax credits). Also, as land reform was at the time under discussion, many landowners cleared the land because this was seen as a proof that it was being used and was thus not subject to potential confiscation for land reform purposes (WRI/UNDP/UNEP, 1990). INPE's 1988 report, however, contributed to highlight the gravity of the situation in Amazonia and to increase the political leverage of deforestation in the international arena. Mounting scientific evidence linking deforestation to climate change highlighted the global impacts that deforestation in Brazil could have for the rest of the world, especially as there had been a particularly dry summer in the USA that year. The combination of these factors heightened public concern with the impacts that deforestation in Amazonia could have on their lives. Finally, in December 1988, Chico Mendes – by now a well known leader of the 'forest people' in the US and Europe – was murdered and this created an international uproar over the Brazilian government's policies in the Amazon region (Goldemberg and Durham, 1990).

International concern with deforestation in Amazonia was not only due to awareness of the global importance of the rainforest in ecological terms. There were factors contributing more than deforestation to the degradation of the global environment, but they received less attention. Burning fossil fuel and cement production, for example, contribute approximately 86% to the greenhouse effect, whereas total world deforestation contributes approximately 14%; deforestation in Amazonia contributes at most 5% (WRI/UNDP/UNEP, 1990). The fate of Amazonia has always had a special place in the concerns of both environmentalists and public opinion in the industrialised countries (Hurrell, 1991; Gross, 1990; Hecht and Cockburn, 1989). In part, this is because Amazonia is the largest remaining continuous tropical forest and, historically, it has always been a myth in the mind of outsiders[13] (Hecht and Cockbun, 1989; Barbosa, 1993). Public concern with Amazonia was also related to the role of the media in presenting the plight of the 'forest people', together with the shocking images of burning forests. Deforestation, as opposed to other environmental issues, 'lent itself to dramatic and extremely effective media presentation. On the one hand, there was the drama and visibility of the process itself, with huge palls of smoke, bulldozers at work, vast areas of the jungle being flooded. On the other there were seemingly clear villains...

and tragic victims' (Hurrell, 1992, p.202). Lobbying against the environmental and social disaster that was seen to be taking place in the rainforest, was also made easier by the fact that this issue provided scope for doing something about it – while not suffering very much at home. Congressmen could lobby for stopping the financing of environmentally damaging projects abroad, and in this way obtain political gains without incurring domestic economic costs (Kolk, 1996).

Notwithstanding the motives for campaigning against deforestation in Amazonia, during the 1980s, local populations saw their 'external context' widen and change: whereas in the 1970s their external context was shaped by mainly national developments, in the 1980s it is shaped by both, national and international developments. The international context also played a role in the 1970s, but its influence on local populations was indirect, affecting them only through the state. In the 1980s international developments had a stronger and more direct impact on local resource users. Paradoxically, one of the reasons why the external context began influencing local populations directly was the shift of focus from the local to the global dimension of environmental problems. As the global implications of deforestation became more apparent, local populations' utilisation of forest resources acquired renewed importance: their use of the resource base, if sustainable, was important for the protection of a globally important resource and not only for their own survival. Hence, the shift from the local to the global dimension of environmental problems provided in many cases political leverage to local populations dependent on natural resources. International environmental agreements that chose to ignore local actors, such as TFAP, were unsuccessful in achieving their objectives. By contrast, the success of the MDB campaign was largely due to the links they established between international and local actors.

The development of environmental concerns in the international arena altered the tappers' external context in at least two ways. First, as highlighted in Chapter 1, the socio-political context of commoners can enhance their capacity to tackle changes in the circumstances if it provides opportunities for commoners developing alliances with other actors; the NGOs strategy in the context of the MDB campaign provided such opportunity, a matter that is discussed in more detail in Chapter 3. Second, the international campaign against deforestation contributed to a considerable extent to change the national setting, especially the government's approach towards Amazonia. This further enhanced the tappers' capacity to deal with the changes in their circumstances that had occurred in the 1980s.

National developments

Brazilian reactions towards international concern with Amazonia

Not all Brazilian sectors of society reacted in the same way to international demonstrations of interest in Amazonia. Some groups, like the government and the military, considered such demonstrations of interest as an interference in Brazilian affairs; others, especially NGOs, welcomed international action against deforestation because this gave them considerable political leverage to change the pattern of development of the region.

A speech by President Sarney in 1989, when international concern with deforestation was particularly vociferous, illustrates the government's initial reaction to international pressure for the conservation of Amazonia: 'Brazil is being threatened over its sovereign right to use its own territory ... With each day there are new forms of intervention containing veiled or explicit threats, designed to force us to take decisions not constructed by us in the defence of our own interests' (quoted in Hurrell 1992, p.405). This antagonistic attitude was based on four major themes, which represented, to a certain extent, the main concerns of the developing world (Hurrell, 1992). One was an emphasis on economic growth, for which the exploitation of Amazonia's mineral reserves and soils was deemed essential. Another was the idea that the North wanted to stop the country from becoming an economic power. Concern with tropical forests was also considered hypocritical, since the Northern countries had destroyed their own forests and grown rich in the process. Finally, the fourth theme was the long long-standing fear that Amazonia is never far from attempts at internationalisation, which was fuelled by both the Brazilian military and international debates about humankind having rights over tropical forests.

Although in the 1980s, the impression that the international community intended to internationalise Amazonia could be found in all sectors of Brazilian society, it was particularly strong among the military, who saw Amazonia as a virtually empty region, with enormous wealth, and therefore vulnerable to foreign occupation. They encouraged foreign investment in the region, but treated any expression of concern with deforestation in Amazonia as an attempt against Brazil's sovereignty over her territory. The military's goal was to transform Brazil into a great power, hence 'all criticism of the [military] government's developmentalist policy necessarily was interpreted as opposition to the regime and was considered a subversive activity that placed national security at risk' (Goldemberg and Durham, 1990, p.31). The military are still highly suspicious of foreigners' concern with the conservation of Amazonia, and although their rule ended in 1985, they have maintained control of some government agencies responsible for Amazonia, such as the Secretariat for Strategic Affairs (SAE). When in the early 1990s Brazil and the G7 negotiated the G7 Pilot Programme to Conserve the Brazilian Rainforests, a programme that has provided considerable support to extractive reserves, fear of international affair in Brazilian affairs was still an issue in the country and it influenced the Pilot Programme negotiations and, indirectly, the extractive reserves' institutional development.

Several occurrences also fuelled the idea that there was a threat to the internationalisation of the region. In the 1960s, for instance, the US Hudson Institute planned to flood large areas of the Amazon region for the creation of lakes. This proposal created considerable commotion in the region and an NGO was formed specifically for the defence of the Amazonia, the National Campaign for the Defence and Development of Amazonia, CNDDA. In the late 1980s, comments made by American senators visiting Amazonia, and President Bush telling the Japanese to stop the financing of BR-364 for environmental reasons, also contributed to the perception that there was a risk of internationalisation of the region. The most referenced comment by a foreign authority on this subject was the speech given by the late French President, François Mitterrand, at the ozone conference in The Hague: he said that an international organisation for dealing with global environmental issues should be created, and that sovereignty should be limited if necessary for the benefit of humankind (Goldemberg and Durham, 1990; Kolk, 1996). Although not explicitly referring to Brazilian Amazonia, Mitterrand's statement illustrated the views of those arguing that, independently of countries' sovereign rights over their territories, humankind has rights over tropical forests.

In academic and policy debates about world deforestation, the idea that humankind has rights over tropical forests because of their global ecological importance is always present (Humphreys, 1996; Thomas, 1992). This argument is interpreted in different ways. Some consider that humankind's rights amount to global responsibility for tropical forests, and, therefore, the international community has the obligation to help cashed strapped governments to conserve their forests. Others go as far as to argue that national states should relinquish part of their sovereignty rights for the common good of all humankind, and that humankind has the right to decide over the fate of forests (Hooker, 1994; Weiss, 1989; Kuehls, 1996; Mishe, 1992). In the South, these arguments are received with considerable animosity and, as it is discussed later in the chapter, they were partly responsible for the little success of negotiations about forests at the Earth Summit in 1992.

The idea that humankind has some sort of rights over tropical forests has two important flaws: first, it assumes that humankind is formed by a homogenous group of people with similar interests; second, it ignores the distribution of power between and within countries. The emphasis on 'people's common interests' steers attention away from 'differentiated social groups and nations having different interests in causing and alleviating environmental problems' (Taylor and Buttel, 1992, p.406)[14]. These differences can be observed between the participants of the MDB campaign, who shared a common interest in forest conservation, but had very different concerns: local groups campaigned for forest conservation in order to maintain their source of livelihood, while environmental NGOs' interest in forest conservation came after their own livelihoods were fully assured. Because of the differences between the two groups, their interest in the conservation of resources may not always coincide. Given a change in the global 'distribution' of problems, international NGOs' support for specific environmental issues can shift in focus; the problems of any local community, however, are unlikely to change in the same direction and, if the original problem has not been solved, the community can suffer from the withdrawal of international NGOs' support.

Although most people in Brazil had encountered at one point or another the thesis of internationalisation, not everybody regarded it as a real danger; instead, many thought that international interest in Amazonia was something positive[15]. In their view, international pressure helped the Brazilian government to construct a new paradigm and speeded up their recognition of the negative impacts their model of development for Amazonia was having. They regarded international pressure to stop deforestation in the same light as they saw Brazilians protests against human rights abuses in other countries, such as apartheid in South Africa. This opinion was very common among NGOs, many of which had, during the 1980s, co-operated with international NGOs. International NGOs co-financed some local projects and national NGOs provided them with the necessary information to undertake lobbying in their countries.

The end of military rule

In 1985, military rule in Brazil ended. This brought many changes to the country, such as the strengthening of civil society groups, a constitutional change and, subsequently, a change in the government's attitude towards development in Amazonia and the international community. The social problems created by the development policies of the previous decades had fuelled the creation of numerous NGOs and grassroots movements, and with the political opening these movements and organisations began voicing their demands more freely[16]. In the same year that the military government stepped down, landless peasants organised a demonstration to demand land reform and, in May 1985, President Sarney presented at the Congress of Rural Workers the new proposal for land reform. Civil groups also helped drafting a new constitution. Established in 1988, the Brazilian constitution is very detailed concerning environmental matters. It includes a whole chapter and a number of sections on the environment, several of which were used later to substantiate the rubber tappers' proposals for extractive reserves. It states that the Amazon rainforest is national patrimony, and that the Government has the duty to protect the flora and fauna of the country, to secure a good environment for its citizens, and to define areas to be preserved (Arnt, 1992; Peixoto, 1996).

In the context of the overall transformation of the political and legal setting of the country, the government's attitude towards Amazonia, the environment, and the international community also underwent considerable change; this also broaden the rubber tappers' range of options concerning the management of their resources. In spite of the continued occurrence of nationalistic speeches, the late 1980s saw a gradual process of shifting the national attitude in relation to the international interest in Amazonia from one of direct antagonism, represented by President Sarney, to one of co-operation, embodied by President Fernando Collor, whose mandate began in 1990. New environmental policies, however, had already begun during Sarney's presidency, and it was during his period in office that extractive reserves were introduced in the legislation as one of the tools of environmental policy[17].

In 1988, President Sarney set up the Brazilian Institute for the Environment (IBAMA), which should formulate, co-ordinate, execute, and enforce the national policy for the environment. Ibama's responsibilities included monitoring fires and illegal clearings, and, upon the creation of extractive reserves in 1990, Ibama was put in charge of administering these institutions in co-operation with the reserves' inhabitants. Like environmental agencies in other parts of the world, Ibama's capacity to monitor the country's environmental policy is insufficient, specially in such a large region as Amazonia: in southern Pará and Acre, Ibama officials have had to request the protection of the federal police to carry on their duties (Schmink and Wood, 1992). Sarney also established the National Environmental Fund (FNMA) with participation of Brazilian NGOs, and on the following year he set up the 'Programme for the Defence the Ecosystem Complex of Legal Amazonia', commonly known as programme 'Our Nature'. 'Our Nature' was a response to the international pressure on Brazil to halt deforestation in Amazonia and, it is often regarded as a nationalistic effort to redefine the nature to be preserved as belonging to Brazilians and not to foreigners (Barbosa, 1993). Nevertheless, 'Our Nature' was also a first attempt at designing a national environmental policy and in this sense was a major step forward (Hall, 1997b).

Fernando Collor began his mandate with a number of environmental measures – considered by some to be more of a facade for international public opinion rather to reflect a real interest in the matter. In a pre-election visit to Europe, Collor had noticed that important economic issues of foreign policy were conditioned by the international criticism of the Brazilian policy for the region (Hurrell, 1992). When he began his mandate he thus 'turned Brazil's tropical forests into a political and economic tool to regain foreign capital. To protect Amazonia, the developed countries would have to share the costs of preservation' (Barbosa, 1993, p.125).

The first environmental initiative of President Collor was to create the Secretariat for the Environment (SEMAM), which was directly connected to the Presidency. SEMAM took the decision-making role of Ibama, and the latter was transformed in the executive agency of the secretariat. To direct the secretariat Collor appointed José Lutzenberger, who besides having given evidence at the US senate in the context of the MDB campaign was an internationally renowned environmentalist open to explore different forms of cooperation with the international community. He examined, for example, proposals by international organisations to reduce the external debt with investments in the environmental area, an initiative that had been strongly opposed by the Sarney government

(Peixoto, 1996). Later, SEMAM was to lead the negotiations on the Pilot Programme, and was the agency responsible for advancing the Extractive Reserve sub-project in the years before the Earth Summit. Collor also set up Operation Amazonia, aimed at stopping fires in the region, and inventorying the wood and metallurgy industries in the region, and the agricultural and mineral projects subsidised by the government. In 1990, the National Programme for the Environment (PNMA) was established; its objective was to strengthen IBAMA and the newly created state environmental agencies (Kolk, 1996; Peixoto, 1996). Fiscal incentives for new cattle ranching projects were stopped and, overall, the attitude of the government towards Amazonia began to change. By 1995, the official government objective for the region was no longer the exploitation of its resources, but sustainable development. This was more than a mere change in discourse. A comment by one of the interviewees encapsulates this change:

> The state is now more responsible in relation to Amazonia and I think that what contributed to that was external pressure, no doubt about that. There is no doubt that Amazonia is today ... the trump of the country in the international arena, a trump for the country's discourse. You can systematically see Brazilian presidents dealing with the environmental issues, when visiting abroad, etc., and they discuss Amazonia. Amazonia cannot be used with impunity because it is a political trump. It has value in international negotiations. In addition, there is now more technical knowledge about Amazonia[18].

International pressure on Brazil was very important in bringing about changes in both the environmental policy of the government and its attitude towards Amazonia. According to Hurrell all other factors promoting change in Brazilian Amazonian policies 'have been over-shadowed by the role of external pressure and by consequent changes in Brazil's calculations of the international costs and benefits of continuing with its previous Amazonian policies' (Hurrell, 1992, p.417). The costs that the campaign against deforestation was having in Brazil were not limited to the hindering of specific projects such as the Polonoroeste: the whole of Brazilian foreign policy was affected. In the closing statement of the seminar 'Amazonia: Facts, Problems and Solutions' (in August 1989, at the University of São Paulo), Ambassador Bernardo Pericas stated that the environmental issue was one of the main diplomatic problems of the country (Peixoto, 1996). As Collor had observed in his pre-electoral visits, economic issues such as debt were being linked to the issue of Amazonian deforestation (Hurrell, 1992). The economic position of Brazil in the international arena – with a large foreign debt – and the fact that the country was having an internal economic crisis further contributed to make Brazil vulnerable to international pressure (Kolk, 1996; Arnt, 1992; Barbosa, 1993). All these factors led to a situation in which Amazonia became for Brazil a political trump in international negotiations and therefore its conservation gained a certain degree of political value.

The global importance of Brazilian Amazonia, both in terms of its ecological value and as a mythical place for people in the North, contributed to change the 'external context' of local resource users in three ways. First, as we saw in the previous section, by expanding the range of developments that influenced them. In the 1950s, since Amazonia was largely unconnected with the rest of the country, it was virtually only what occurred at the regional level that affected local communities; in the 1970s, because of the government policies for the region, the national political and legal setting began impacting on resource users and; by the late 1980s, they were influenced by developments at the international level. Second, because of the global importance of Amazonia, the range of actors with whom resource users in the region could develop alliances and obtain support increased. This, as discussed in the first chapter, facilitates resources users' adaptation to new situations. Third, the global importance of Amazonia led to an international campaign that triggered or speeded up a change in the government's attitude towards the region. The new legal, institutional and political setting of Brazil provided a new range of opportunities for the tappers' struggle against the cattle ranchers, as will be demonstrated in Chapters 3 and 4.

The Earth Summit

The growing interest in environmental matters observed during the 1980s reached a peak in June 1992, at the United Nations Conference on Environment and Development (UNCED) or the Earth Summit. Most analysts regard the results of this conference, in terms of documents agreed upon, unspectacular (Grubb et al, 1993; Johnson, 1993; Kolk, 1996). Nevertheless, the Earth Summit had an unparalleled role in temporarily bringing environmental issues, especially those of global importance such as deforestation, to the forefront of international political affairs. In the context of this study, the importance of UNCED may be pinned down to two issues: first, the anticipation of the conference induced Brazil and other countries to take a number of environmental measures, some of which brought direct benefits to the rubber tappers; second, participants at the conference almost unanimously agreed that the conservation of forests depended on local populations' sustainable use of their resources. Forest conservation and international co-operation on the matter was the most controversial issue at the Earth Summit, and consensus on the role of local people may have occurred because support for local communities can circumvent to a certain extent disagreement on whether national governments or the 'international community' should have most rights over tropical forests. The implication of such consensus, however, for local resource users, is that their potential political leverage can increase and with it their chances of obtaining external support. The next two chapters argue that this occurred with the rubber tappers.

The Earth Summit took place in Rio de Janeiro, during the mandate of Fernando Collor. The offer to host the conference, however, had been an initiative of President Sarney to demonstrate interest in the conservation of the global environment. This could help to reduce the international pressure on the country's destruction of its forests. In addition, hosting the conference would increase the bargaining power of Brazil, and change the perception of rich countries concerning environmental problems in the developing world (Peixoto, 1996). During the run up to UNCED, Collor established 10 environmental units, including five extractive reserves, and signed a programme of co-operation with the G7 countries to promote sustainable development in Amazonia, the Pilot Programme for the Protection of Brazilian Rainforests (Peixoto, 1996). This programme includes among other sub-projects one specifically aimed at supporting extractive reserves, and it was in the run up to the conference that the procedures for the implementation of his sub-project were most advanced.

The position of the Brazilian government during the preparatory process to the Summit, and during the conference itself, was based on two points. The first was the acknowledgement that global environmental problems are important, and that they must be dealt with mainly through international cooperation. This represented a more co-operative attitude than the one of the military and of President Sarney in previous years, when environmental concerns with deforestation were regarded as disguised attempts at the internationalisation of the region. The second point was that there is differentiated responsibility for the cause and correspondent solution to environmental problems, and so rich countries must bear a higher cost (Viola, 1993).

Amazonia was not a specific topic of discussion at UNCED[19]. All the same, and in spite of the more co-operative approach of Brazil during the Earth Summit, the country strongly rejected the North's views on forests, as did most of the South[20]. In the Preparatory Committee meetings for the Conference (prepcoms) a number of North-based organisations had proposed a global convention on forests, but the South strongly opposed this proposal, which they saw as interfering with their sovereignty[21]. They stressed instead the need for funds and technology to preserve their rainforests. The Brazilian Minister of the Environment also stated that Brazil would reject unilateral measures to protect forests that could harm the Brazilian economy. Besides, before signing a forest convention it was necessary to resolve the issue of greenhouse gases (Kolk, 1996). The suggestion for a global convention was dropped during the prepcoms, but the contention between the North and the South regarding forests continued throughout the conference.

Broadly speaking, the North viewed tropical forests as a common concern of humankind, and, therefore, it thought that the international community should have a say in their management. For the North, a convention on forests was a form of co-operation among various interested states, whereas for the South such a convention was an attempt to exert supranational control (Kolk, 1996). The South linked the issue of forest conservation to the inequality existing between the two groups. Developing countries pointed to the historical responsibility of the industrial world in greenhouse emissions and how this was partly accountable for their level of development. They stressed their own need for development and the need for financial transfers from the richer countries if they had interest in the conservation of developing countries' forests. The Southern states also mentioned the fact that industrialised countries were still responsible for most greenhouse emissions, and they accused the North of putting more emphasis on tropical forests conservation rather than on the reduction of their own emissions of greenhouse gases. Furthermore, the South considered the issue of tropical deforestation as related to the present structure of the economic system, the issue of debt and the role of these factors in promoting deforestation. Although the proposal for a convention on tropical forests had been dropped, the issue of sovereignty was nevertheless particularly strong, since the South would not agree 'to be told what to do with their forests even if they were offered compensation by the North' (Sullivan, 1993, p.161). The South also disagreed with the proposal of the North for an agreement on tropical forests only and argued for the inclusion of all forests in the agreement.

In the end, participants at the Earth Summit signed the Authoritative Statement on Forest Principles, which is a non-binding document and includes all types of forests, including those located in temperate regions of the North. The Statement, according to most analysts, represents the lowest common denominator between the objectives of the North and the demands of the South (Grubb et al, 1993; Johnson, 1993; Humphreys, 1996). While the North failed to obtain its desired global convention, the South did not obtain the financial commitments from the industrialised countries it had demanded. The global value of the forest is noted in the final document, but with less strength than what was originally intended by the proponents of a global convention; on the other hand, although the linkage between deforestation and the global economic system is mentioned, there is no acknowledgement in the document of developed countries' responsibility. The text recognises the need for international co-operation, but there are no specific prescriptions for future collaboration on forest issues (Humphreys, 1996; Sullivan, 1993; Kolk, 1996; Johnston, 1993).

The Statement on Forest Principles represents the state of global consensus in relation to forests that existed at the time, and although the document is non-binding, it can be used as a yardstick to judge governments' policies towards forests. The Document states in the Preamble that forests are valuable for local populations. Later, there are references to the need to integrate local communities in the 'development, implementation and planning of national forest policies' (Section 2d). Support for local communities should come from both national governments and the international community: 'The problems that hinder efforts to attain the conservation and sustainable use of forest resources and that stem from the lack of alternative options available to local communities, in particular the urban poor and poor rural populations who are economically and socially dependent on forests and forests resources, should be addressed by Governments and the international community' (Section 9b). The role of the international community, however, can be assumed to be subordinated to the states' 'sovereign and inalienable right to utilise, manage and develop their forests in accordance with their development needs and level of socio-economic development...'

previously stated in Section 2a, and, in different phrasings, in various other parts of the document (Johnson, 1993).

The Statement also acknowledges the importance of land tenure systems in both the conservation of forests and the well being of local users. 'Appropriate conditions should be promoted for these groups to enable them to have an economic stake in forest use, perform economic activities, and achieve and maintain cultural identity and social organisations, as well as adequate levels of livelihood and well-being, through, *inter alia*, those land tenure arrangements which serve as incentives for the sustainable management of forests' (Section 5a). Hence, besides acknowledging the existence of different property rights systems, the Forest Principles also indicate that the North and South agree on the need to support such systems, if they contribute to the sustainable use of resources. The content of the Statement on Forest Principles suggest that common property institutions, which until recently had been associated with resource depletion, are now regarded as capable of securing the conservation of forests. It also suggests that the North and the South agree that external agencies should not interfere with local users' organisational arrangements. The Forest Principles can thus be considered to be facilitative to co-owners of natural resources because it is a legal item that recognises the potential of their property rights institutions to ensure resource conservation.

There are also references to forests in Agenda 21, another document agreed on in Rio whose aim is to set out an international programme of action for achieving sustainable development in the 21st Century (Johnson, 1993). Chapter 11 ('Combating Deforestation') of Agenda 21 deals with forests. It lists four objectives and specifies the basis of action, objectives, activities and means of implementation relevant to each of them. Land tenure is mentioned as one of the bases for actions in relation to Objective B, 'Enhancing the protection, sustainable management and conservation of all forests, and the greening of degraded areas, through forest rehabilitation afforestation, reforestation and other rehabilitative means'. This item states that action for conserving and sustaining forest resources 'should include the consideration of land use and tenure patterns and local needs ...' but there are no further references to the relationship of land tenure to sustainable forest use[22].

In sum, the Earth Summit produced legal documents that are important for local forest users because of the three reasons. First, they recognise the right of local peoples to their forests; this is important because local people have often been expelled from their areas for the sake of conservation. Second, they acknowledge the importance of local people in the solution of globally important problems; this can increase their political leverage, and their capacity to establish alliances with external actors. Third, they recognise that private and state property regimes are not the only forms of property institutions there are, or the only ones that can ensure sustainable use of resources; this further enhances the possibility of local peoples to see their rights recognised.

The Conference was a landmark in the process of international co-operation on environmental matters and it triggered a number of environmental initiatives. Once the conference was over, however, the environmental drive of the previous years slowed down. Both governments and the general public lost much of their interest, and even Amazonia, which had been such a hot topic in the previous years, lost priority. Although non-governmental organisation continued campaigning against deforestation in Brazilian Amazonia more than about deforestation in other regions, the Amazon rainforest never regained the centre stage position of the late 1980s. By 1999, reports showed that deforestation was higher than it had been during that period, but this grabbed little attention outside restricted circles.

The loss of interest in environmental matters in the aftermath of the Earth Summit was partly due to a certain fatigue among participants resulting from the high expectations and the minimal results (Kolk, 1996). Some events in Brazil also contributed to the loss of political leverage of Amazonia and environmental issues. Shortly after the Earth Summit, Brazil went through a political crisis that culminated in the impeachment of President Fernando Collor in September 1992. The recession that the country went through during that period also contributed to diminish the newly developed concern with Amazonia (and to decrease deforestation in the region as well). At the international level, the issue of Amazonian deforestation lost some of its salience given the prevalence of other issues such as the collapse of the Eastern/Socialist bloc. Nevertheless, the Earth Summit represented a turning point concerning the attitude of the international community in relation to Amazonia: from the Summit onwards the region began to be seen more in need of sustainable development than as either a conservation shrine or an exploitable resource.

As it will be shown later, this change of attitude about Amazonia can be of particular importance for local populations who depend on forest resources. If environmental conservation is at the centre of attention, forest users can more easily obtain external support (financial or otherwise) than if the public ignores their contribution to society. Nevertheless, once a process is set into motion, it may continue even after the original trigger has lost strength. This occurred with the rubber tappers: global concern with deforestation in the 1980s triggered a number of actions, including a programme for the support of the forests, the PP-G7, which have continued to influence their lives and institutions long after their plight has left the newspapers of the North.

The G7 – Pilot Programme to Conserve the Brazilian Rainforests

The G7 – Pilot Programme to Conserve the Brazilian Rainforests (PP-G7) is one of the most important initiatives resulting from the national and international developments reviewed in the previous pages. First, it represents a change from the initially widespread idea that Brazil had an obligation to protect its forests, to the view which emerged in the late 1980s 'that Brazil should probably be given funds on favourable conditions if the industrial countries attached so much importance to the conservation of Brazilian rainforests' (Kolk, 1996, p.146). Second, it involves a new approach to the region from both the Brazilian government and the international community, in which local people have more prominence. The PP-G7 is of utmost importance for extractive reserves, and it has played a noticeable role in shaping the rubber tappers' institutions. The level and form of support, however, has been shaped by a number of international developments – the PP-G7 is therefore a case in point of how the global political economy can influence local resource users.

The proposal originated in 1990 at the Houston Summit of the Group of the seven most industrialised countries (G7), where Germany suggested setting up a pilot programme to help Brazil conserve its rainforests (Hagemann, 1994; Kolk, 1996). The Brazilian government accepted the offer and set up, shortly afterwards, an inter-ministerial commission to draw up a proposal. Within the government, however, views about the Pilot Programme diverged: a group led by SEMAM supported it, whereas the Foreign Ministry, Itamaraty, had several reservations about it. Itamaraty feared that the programme could involve international interference in areas of national interest, such as occupation of forestland and road construction

(Hagemann, 1994). Itamaraty and its allies in the government proposed that Brazil should compile a proposal made of projects already budgeted. Secretariat for the Environment, on the contrary, argued that the government should present a proposal made of new projects, and that the programme should favour groups in the Amazon that had never benefited before from development plans. This was, for instance, the case of the rubber tappers and other local populations such as the river dwellers (*ribeirinhos*) who all had been ignored in the previous government policies for the country. In SEMAM's opinion, local communities could play an active role in the implementation of the programme: 'Local communities were to be enabled to carry out their own projects, demonstrating to society the feasibility of sustainable development and encouraging a change of attitude towards deforestation' (Hagemann, 1994, p.71). SEMAM thus considered that extractive reserves should be supported in the context of the PP-G7, and considerably advanced this component of the programme.

Several factors helped to balance the dispute towards the acceptance of the Programme and to favour SEMAM's proposals (Hagemann, 1994; Kolk, 1996; Fatheuer, 1994). One was the anticipation of the Earth Summit. Co-operating with the industrialised countries in an initiative aimed at protecting the forest would give a positive image of the country, one of Collor's aims. Furthermore, the idea that the environmental cause could yield substantial international funds was already part of the new government's strategy, and the Pilot Programme seemed to exemplify this. Given the political context of the early 1990s, SEMAM's views on the programme were favoured over those of Itamaraty, and hence support for the rubber tappers, an issue that donors also favoured, was already part of the first version of the PP-G7.

Almost from the beginning, NGOs had a bearing on the PP-G7. NGOs, especially Friends of the Earth (FoE), had been lobbying the G7 since their 1989 Summit. The Instituto de Estudos Amazônicos e Ambientais (IEA), one of the main rubber tappers' supporting NGOs, was in close contact with FoE, and was well informed about the Programme from an early stage (Hagemann, 1994). In 1991, the Brazilian government invited Brazilian NGOs to participate in the design of the Programme.

Among Brazilian NGOs there were two positions in relation to the Pilot Programme. One group, namely the Brazilian Forum of NGOs, strongly opposed the Programme because public participation in its design had been supposedly insufficient. Forum (formed by over 700 members) argued that the rationale and content of the approach had not been developed by

Brazilian society, but rather by the international community. Consequently, the programme took an ecological perspective of the Amazon region, instead of considering the social diversity of Amazonia. The other group of NGOs had the same reservations as Forum about the programme, but decided to work together with the governmental parties in the reformulation and implementation of the PP-G7. This second group of NGOs (more than 200) constituted the Grupo de Trabalho Amazônico (GTA) to represent the non-governmental sector at the PP-G7 negotiations (Fatheuer, 1994; Hagemann, 1994).

The Pilot Programme's objective is embedded in the new concept of sustainable development, in the sense that it aims at conserving the environment through development, and at developing without destroying the natural resource base. Yet, the central purpose of the programme is the conservation of the global environment rather than development per se; the latter receives support only insofar as it contributes to the conservation of natural resources valuable for humankind. The programme's documents state that the overall objective of the PP-G7 is 'to maximise the environment benefits of Brazil's rainforest consistent with Brazil's development goals, through the implementation of a sustainable development approach that will contribute to a continuing reduction in the rate of deforestation' (WB/CEC/GoB, 1991, p.I). Its specific objectives are 'i) to demonstrate the feasibility of harmonising economic and environmental objectives in tropical rainforests; ii) help preserve the huge genetic resources of the rainforest; iii) reduce the Amazon's contribution to global carbon emissions; and iv) provide another example of co-operation between developed and developing countries on global environmental issues' (WB/CEC, 1991, p.3).

The objectives of the PP-G7 are to be achieved through four sub-programmes. Three are 'structural projects' aimed at improving environmental institutions at the national and regional level, and one is formed by 'demonstration projects' that should be proposed and implemented by both citizens' groups and governmental agencies (FoE/CTA, 1994).

Funding for the Programme was less than anticipated. The donors had not given explicit indications of how much funding would be available, but a programme for the whole of Amazonia and the Atlantic Rainforest, financed by the G7, could expect a budget in the order of a billion dollars; programmes with similar features but with only regional or more restricted objectives, e.g. the National Programme for the Environment and PMACI (the programme for mitigating the negative effects of the POLONOROESTE), had budgets of one or two hundred million US dollars (Hagemann, 1994).

Brazil's initial proposal for the PP-G7 totalled US$1.566 billion for a five-year period, but by 1991 the agreement was that the donors would provide approximately US$250 million for the 3-year initial phase (Hagemann, 1994). Approximately only half of these funds consisted of new donations; the rest was for already planned projects that were fully consistent with the PP-G7. Most of the funding was bilateral. The Forest Trust Fund, a core fund for multilateral contributions administered by the World Bank, had in March 1992 a size of US$60 million. In sum '[m]ost commitments were relatively small, with minimal or no contributions to the multilateral fund and a larger share to already planned projects and new bilateral funds. A significant portion of the latter funds was already earmarked for prepared or appraised projects' (Hagemann, 1994, p.117).

Several factors can explain the limited financial commitment of the donors when compared with initial expectations. First, several countries, especially Germany, Canada and the UK, stated that funds could only be allocated on a bilateral basis. Second, there were political differences among the several parties involved. Some countries fully supported the programme (Germany, Italy, Canada and the UK) whereas others had certain reservations (Japan and the US) and considered that the project lacked maturity (Hagemann, 1994, p.105-106). In addition, in 1991, the environmental issue was losing some of its momentum in face of the difficult situation in the Soviet Union and Eastern Europe, which required additional funds (Kolk, 1996).

Although the programme was approved by the end of 1991, its implementation began only in 1994-1995. This delay, which slowed down the legalisation of the rubber tappers' property rights institutions, was due to both international and national developments. As noted earlier, concern with the environment and Amazonia declined after the Earth Summit – the G7 interest in the programme declined accordingly. In Brazil, at around the same time, there had been a change in government (because of President Collor's impeachment) and the new Secretary of State for the Environment, a former diplomat, thought that the PP-G7 focused too much on Indians and Extractivists (such as the rubber tappers), and announce that neither of the two would be priorities for the new government (Hagemann, 1994).

Table 2.1 PP-G7 sub-programmes and their components

Natural Resources Policy	Conservation and Natural Resource Management Units	Natural Resource Management	Demonstration Projects
Economic-ecological zoning	Parks and reserves	Recovery of degraded areas	Type A demonstration projects
Environmental monitoring and surveillance	National forests	Science and technology	
Environmental control and inspection	Extractive reserves	Centres of excellence	
Institutional strengthening of state environmental agencies	Indigenous reserves	Directed research	
Environmental education			

Source: FoE/CTA, 1994.

There are four main criticisms to the PP-G7: first, the programme sees Amazonia mainly as forest, and ignores the social complexity of the region; second, it fails to address the underlying causes of deforestation; third, there is lack of articulation between the PP-G7 and other national policies for the region; and fourth, the implementation of the programme has been very problematic.

The PP-G7 supports Amazonia's traditional inhabitants or 'forest people', such as Indians and rubber tappers, but it largely ignores the urban population (which represents over 50% of the inhabitants of the region), and the migrants and colonists who arrived in the last 30 years (FASE/IBASE, 1993; Fatheuer, 1994). Support for the region's traditional inhabitants is, however, a positive development. Whereas in the 1970s and 1980s, the development model for Amazonia was based on mega-projects and the opening of highways, the development model represented by the PP-G7 focuses on forms of production that are more adequate to the ecological characteristics of the forest and that benefit a sector of the

Amazonian population that undoubtedly needs support. '[The programme's] importance lies in the fact that the PP-G7 has been able to support a number of pre-existing institutions and initiatives' including, apart from extractive reserves and demarcation of indigenous lands, research institutions in Amazonia and state environmental monitoring as well as 'encourage innovative action on a number of other fronts' (Hall, 1997a, p.66)[23].

One of the main underlying causes of deforestation is the land tenure situation in Amazonia, which is ignored by the Programme (FoE, 1991). Failing to address this issue, however, 'may lead to the creation of 'islands of conservation' [...] the major stimuli to deforestation will not be tackled; namely, land concentration and social pressures in other areas of Brazil which encourage migration' (Hall, 1993, p.9) With regard to lack of articulation of the PP-G7 with other government policies that indirectly lead to more deforestation, one of them is the rubber policy: on the one hand, the PP-G7 supports four extractive reserves, but on the other hand, the governments' rubber policy has seriously undermined the tappers' capacity to make a living in the reserves.

The implementation of the programme has been problematic because of several reasons. For instance, as each of the projects composing the overall programme is negotiated separately, and often in the context of bilateral aid mechanisms, those issues which are more attractive to the public in the industrialised countries may receive most of the funding (Batmanian, 1994). The rubber tappers, an 'attractive' issue in the early 1990s, may have benefited from this bias. The bureaucracy of the Brazilian government has also contributed to slow down the project implementation (Fatheuer, 1994), a criticism that has been voiced in relation to the specific case of the extractive reserves. The participation of NGOs, although generally agreed to be one of the important positive aspects of the Pilot Programme, has also been limited by their insufficient technical capacity – an issue that the PP-G7 has failed to address (Fatheuer, 1994; FoE/GTA, 1994).

Although the programme has not been as comprehensive as expected and has failed to address crucial issues, by and large its impact in the region has been positive. It has helped to reshape the Brazilian government policy for the region (Hall, 1997a). It has also supported local populations that were until recently ignore by government policies. Moreover, it has encouraged the development of alternative economic alternatives for Amazonia. Subsequent chapters examine the influence that the programme has had on the rubber tappers' institutions, and highlight some of its shortcomings as far as the extractive reserves' inhabitants are concerned.

Conclusion

Chapter 1 argued that developments in the external context can trigger changes in the commoners' circumstances, changes to which the resource co-owners have to adapt to prevent the depletion of the CPR. The capacity of commoners to adapt to a change in circumstances depends on internal factors (e.g. characteristics of the resource and the group of users) and on the external context. The external setting in which resource users are embedded, such as the socio-political and economic context, the legal setting and the role of the state in relation to common property regimes and related issues, can thus be more or less facilitative in relation to common property regimes. This chapter has presented an overview of the tappers' external setting from the 19th century to the early 1990s, focusing on the period between the 1970s and the 1980s. The next three chapters will examine the process of development and characteristics of extractive reserves in the context of the developments reviewed in the present chapter. Before doing this, however, it may be useful to summarise what these developments were.

Until the 1970s, the inaccessibility of the region, and the fact that land was not integrated in a market economic system shaped the specific features of the property rights systems of the Amazon. During the rubber boom, another important external influence was the international demand for rubber. After the collapse of the rubber trade, however, and since the state was largely indifferent to the region, it can be assumed that common property regimes had to deal mainly with internal factors, such as the features of the CPR and the characteristics of the users.

In the 1970s, the government approach to the Amazon region triggered changes in the circumstances that, according to the theory on common property, can threaten the boundaries established in the context of common property regimes. Highways were opened across the Amazon, migration to the region was encouraged, and the state catalysed investment in ranching, mining, and large-scale agriculture. These policies attracted vast numbers of people to Amazonia. The rise in the demand for land, which was now part of the national land markets, combined with insecure property rights and the promotion of unsustainable activities, gave rise to a major conflict over rights and to high levels of deforestation. During the 1960s, 1970s and early 1980s, the main actor in Amazonia was the national government, which 'saw the region almost exclusively as an infinite resource pool which could be tapped at little or no social or environmental cost to serve a range of economic, strategic and political interests' (Hall, 1997b, p.61).

Another change in the circumstances that can alter the incentives of resource users is sudden access to a market economy. Although Amazonians had always participated in a market economy for the commercialisation of their produce (especially rubber), the development of a market for land was a new development.

With regard to the influence of the external context in enhancing or hindering the capacity of joint users to deal with the referred changes in circumstances, the legal setting was not facilitative, since it ignored the existence of common property. Moreover, the state failed to secure law and order in the region, which is likely to have increased the difficulties that commoners faced in relation to non-owners.

This chapter has also shown how the arguments as to why resource depletion occurs, which were outlined in Chapter 1, are applicable to the Amazon region. The theoretical framework highlighted the following situations as endangering the conservation of the resource. First, if the resource is in an open access situation and there is scarcity, in which case users are likely to extract as much as possible from the resource in the short term, as they have no security of being able to do so in the future. Second, if the resource is held as state property, but the government lacks the capacity to secure the sustainable use of the CPR, or is uninterested in environmental issues. Third, if the resource is privately owned and it is the efficient decision for the private owner to overuse the CPR.

These three situations could be observed in the Brazilian Amazonia. As there was no law and order to secure the informal rights of the traditional populations and of many of the peasants who migrated to Amazonia, property rights in the region were unprotected, which meant that there was *de facto* an open access situation. No one had secure rights to the land and clearing the forest was the best means of securing those rights, either for speculative motives or to obtain compensation for being expelled from the land. Moreover, the state failed to ensure the sustainable use of the forest because conservation of natural resources was not part of its objectives. In fact, rather than setting up policies that encourage the sustainable use of forest resources, it set up incentives that encouraged the destruction of the natural resource base. A case in point was that landed property rights were granted to those that had cleared the land. And, for private owners clearing the forest cover was the most efficient solution, because the rate of return on cleared land was higher than that of forested land.

According to Chapter 1, a fourth situation that can lead to resource depletion is if a resource is held in common, but co-owners are unable to adapt to a change in the circumstances that increases their own incentives to free-ride, encourages non-owners to use the CPR, or does both. The developments that occurred in Amazonia during the 1970s created a change in the circumstances for all commoners living in the region. The remainder of the book examines how a particular group of commoners in the Amazon region, the rubber tappers, dealt with such changes.

The present chapter identified three reasons why, when examining common property rights institutions in Amazonia, it is necessary to include international factors in the analysis: the global ecological importance of Amazonia, the perceptions of the general public about it, and the global political economy. The forest plays an important role in the global climate and harbours a rich variety of species, which are valuable for humankind as a whole. The high levels of deforestation that have occurred in certain states of Amazonia are thus harmful not only for populations who depend on the forest, but also for the international community. Throughout the 1970s, 1980s and 1990s, the importance of Amazonia in ecological terms has, however, remained the same, but it was only in the 1980s that the international community intervened in the region to halt further deforestation – this was because it was in the 1980s that the global ecological importance of Amazonia became widely known. In the 1970s, international influence in the region occurred mainly because of economic imperatives, and later the international community had so much influence in the region partly because of the economic position of Brazil in the world. In the 1980s, it was international concern with deforestation that played the most important role; accordingly, the main reason why external factors influenced the 1980s was that their common-pool resources were part of a globally important resource.

Whereas in the 1970s, the role of international actors in relation to Amazonia was virtually limited to support the Brazilian government policies for the region, in the 1980s international actors NGOs in particular, took an active role in altering the government's approach to the region and promoting a different set of policies for Amazonia; in this context, they seek alliances with local populations. There is a change in the general international socio-political context. Environmental issues acquire prominence outside restricted circles of scientists and environmentalists, the global ecological importance of Amazonia becomes common knowledge and the term 'forest people' becomes associated with the sustainable use of forests resources. These issues are discussed in the media and at the intergovernmental level. The issue of deforestation in Amazonia, for example, was brought up in meetings between Collor and foreign heads of state, American senators travelled to Amazonia, and President Mitterrand outlined his view on the measures that may be necessary to implement to secure the conservation of globally important resources.

These changes in the international socio-political context were mirrored in the international legal setting. At the Earth Summit, apart from the conventions on biodiversity and global climate change (where forests have a role) there was also a particular document on forests. The Forests Principles and Agenda 21, highlight the value of forests for populations who depend on them for their survival, their need for support from both national and international bodies, and the importance of their land tenure arrangements in the sustainable use of forests. By the end of the decade, there were indications that there was also a change (albeit not radical) in economic terms with regard to Amazonia. Loans for projects in the region started considering potential environmental impacts, and industrialised countries began to consider supplying financial resources to promote the conservation of the rainforest. It was in this context that the proposal for the PP-G7 originated.

The national socio-political context also changed during the 1980s. With the end of the military government, local groups became able to voice their demands more freely, civil society was able to participate in political decisions, and environmental issues acquired a higher political profile. The new 1988 Constitution reflects many of these changes; environmental issues, for example, receive wider coverage than in the previous constitution. The government develops a new approach towards both the environment and Amazonia. A number of new institutions are created with the specific aim of dealing with environmental issues and with the Amazon in particular, such as Ibama, SEMAM, Operation Amazonia and Our Nature. The changes in the government approach to Amazonia are also related to economic issues. Some of the projects in Amazonia where partially financed by the MDBs, and this made it possible for environmental organisations in that country to press for changes in the Brazilian government's approach to Amazonia. The capacity of international actors to influence the Brazilian government policies for the Amazon, was also influenced by Brazil's need to negotiate debt issues and future international loans with foreign governments.

Within the overall trend of the decade of more interest with the conservation of the Amazon rainforest, some developments provoked changes in the political profile of local populations in the region. The anticipation of the Earth Summit served as a catalyst for environmental initiatives, by both the Brazilian government and by industrialised countries' governments, which included support for the rubber tappers. With the end of the Summit, the interest in environmental issues and by extension in the rubber tappers diminished. The impeachment of President Collor and the consequential change of government shifted the balance of power between different government agencies, and this in turn changed the interest of the government in supporting rubber tappers. The recession that Brazil went through shortly after the Earth Summit was also a factor arising from the economic setting which influenced the amount of support that environmental projects and local population received during that period.

From the 1920s to the 1990s, the external context of the rubber tappers thus evolved from 'indifferent', to hostile, and to facilitative. The subsequent three chapters explore how these changes in the external context interacted with internal factors pertaining to characteristics of the tappers and their forest, and influenced the development and characteristics of extractive reserves.

Notes

[1] The figures presented here refer to Amazonia Legal, an administrative region (created in the 1960s) covering an area of 5 million sq. km. and constituted by seven states: Acre, Rondônia, Amazonas, Pará, Amapá, Roraima, Mato Grosso, part of Tocantins and Maranhão. Amazonia Legal does not correspond precisely to the Amazonian ecosystems, which form only 74% of it, but it is the definition of Amazonia that official statistics as well as most analysts use.

[2] The Amazon soils are very poor because most of them are of very old geological origin and many of the essential nutrients have been leached due to the impact of heavy rains and high temperatures. The soils in the region have also high concentrations of aluminium and hydrogen and thus their capacity for retaining nutrients from decomposing matter is low; in addition, the soils' high acidity reduces the availability of nutrients to the plants (Jordan, 1985).

[3] A tight nutrient cycle prevents any nutrient loss from the system. The thick canopy tree absorbs all nutrients dissolved in rainwater and from atmospheric particles, while at the ground level, there is a large and diverse number of organisms, fauna and litter communities, that recycle the nutrients from dead plants and animals (Fearnside, 1986). The roots of trees, which form a dense layer at the surface of the floor, absorb nutrients directly from the litter, in this way avoiding the possibility of nutrients left in store in the ground being lost (Junk and Furch, 1985; Furley, 1990). The above ground biomass also reduces the impact of rainfall, that could carry away the nutrients, retains the moisture and absorbs and conserves the nutrients; the remaining nutrients are conserved by mycorrhizae, a type of fungus (Sioli, 1985).

[4] Erosion takes away the superficial layers of the soil, which in the Amazon, as opposed to other regions of the world, are more fertile than the deeper ones (Fearnside, 1985; Sioli, 1985; Herrera, 1985).

[5] Each greenhouse gas has a different warming potential, according to its atmospheric lifetime and capacity to absorb infrared radiation (WRI/UNDP/UNEP, 1994).

[6] Although the role of deforestation in contributing to the greenhouse effect can not be underestimated, it needs to be put in context. While deforestation in Brazilian Amazonia in the 1980s contributed 4-5% to the total of CO_2 emissions, the US in 1987 contributed 17.6% of total emissions, followed by the USSR with 12%. In the third place came Brazil, considering the country's emissions from industrial processes as well as from deforestation in the Amazon region (WRI/UNDP/UNEP, 1990).

[7] Whereas in most areas of the world rain is mainly generated through vapour coming from the ocean, in the Amazon rainforest over 50% of the rain come from evapotranspiration from the forest (Salati, 1985; Fearnside, 1985). Deforestation, by changing the leaf structure of the vegetation responsible for the process of evapotranspiration, is thus likely to have a considerable impact in the rainwater levels of the region. Moreover, deforestation destroys the root mat in the surface of the floor, which absorbs a high proportion of water that can be recycled in the atmosphere through evapotranspiration (Sioli, 1985). As the bare soil absorbs water at much lower rates than under forest cover – e.g. under pasture conditions they absorb water at less than a tenth of the rate (Lovejoy, 1985) – clearing the forest cover increases the amount of water which is recycled in the atmosphere.

[8] A large reduction of the forest cover would lower heat absorption, decrease evapotranspiration and diminish the heat flux; all of these can weaken global air circulation and thus reduce rainfall (Fearnside, 1985). Others scholars have mentioned the possibility of global implications 'deriving from reduced cloud formation. Smaller amounts of water vapour must be lifted higher before a cloud will form, and cloud formation provides a major form of heat transfer, thereby playing a significant role in global heat balance' (Lovejoy and Salati, 1983, p.215).

[9] There are between 5 and 10 million plant and animal species on Earth, and of this number one tenth is believed to be in the Amazonia (Browder, 1989). Other studies estimate that the global number of species may be as high as 30 million, and that more than half of this biota is in the Amazon region (CDEA, 1992). Already identified, there are 60,000 plants and 2 million insects and microscopic forms of life in the rain forest (De Onis, 1992). There are several explanations for the extraordinary biodiversity of Amazonia. One is that distance between members of the same species helps to insulate against pests and diseases, more easily spread in hot climates (Janzen, 1970; Fearnside, 1986). Another explanation is that the existence of a large number of species is a device to cope with the poverty of the Amazon soils: different species have different nutrient requirements and different capacities for absorbing the small quantities of available nutrients (Junk and Furch, 1985).

[10] This rich variety of species can be easily destroyed for a number of factors, in particular their endemism: most species are limited to particular areas, and they only exist in specific ecosystems within the rainforest. Furthermore, they can not live in isolation: although they tend to be concentrated in small clusters they need large extensions of forest and the existence of other species, in order to be able to perform the necessary ecological functions, such as polinisation, and to have protection against pests and diseases. The latter are very common in Amazonia because of the humidity and heat of the region, and the lack of a winter season as in temperate regions (Fearnside, 1986, 1990). Apart from the endemism, some regions of the Amazonian rainforests have clusters with very large number of species. The clusters or *refugia* are small areas with a very large number of endemic species, limited in distribution (Lovejoy and Salati, 1983). Both the endemism and the existence of clusters means that large numbers of unique species can disappear with relatively small levels of deforestation (Fearnside, 1990).

[11] The concept of sustainable development draws on similar ideas formulated by the International Union for the Conservation of Nature and Natural Resources in 1980, and on studies conducted by several researchers in the 1970s (Kolk, 1996).

[12] For different interpretations and analyses of the concept see, for example, Pearce et al (1989), who present a neo-classical view on sustainable development; Redclift (1992) and Adams (1990), who argue that the concept is human based, and Lélé (1991), who criticises the contradictions of the concept.

[13] The immensity of the region, the dangers of entering it, as well as the secrecy about the region kept by both Portuguese and Brazilian governments, contributed to make of the Amazonia a myth of immensity and mystery. Initially, upon its 'discovery' by the Portuguese, Amazonia was a land of wealth, the El Dorado; it was also the 'green hell' and, at various points through history but specially during the last 20 years, it has been the 'lost Eden', the realm of nature.

[14] For an interesting examination of this issue from a moral perspective see McCleary, 1991.
[15] For more detailed analyses on the issue of internationalisation see Miyamoto (1989), Kolk (1996), de Assis Costa (1990) and Goldemberg and Durham (1990).
[16] These organisations had developed out of the need to support local groups, such as the tappers, through technical, financial and educational assistance. Many of these organisations came out of the Catholic Church, such as the Pastoral Land Commission (*Comissão Pastoral da Terra*) which regularly reports on violence in rural areas, which in the large majority of cases is related to disputes over landed property rights. Some organisations also had a clear political agenda, such as the Rural Unions. Apart from support organisations, also many grassroots movements formed in the region, the most well known being the Survivors of the TransAmazônica, the Collectors of Babaçu and the Rubber Tappers' Council. For more information on this issue see Schmink and Wood, 1992, and Porro, 1995.
[17] For a review of the evolution of Brazil's environmental policy before the 1980s see Hall (1997b, p.53-55).
[18] O poder público é muito mais responsável. E acho que até contribuiu para essa responsiblidade a pressão estrangeira, não tem dúvida. Assim como não tem dúvida como a Amazônia hoje em dia, ..., é o triunfo do país nos forums internacionais, é um triunfo inclusive para o discurso do país. Você vê sistematicamente presidentes brasileiros tratar da questão ambiental, quando fazem visitas, etc e tratam da Amazônia. Ela não pode ser usada impunemente porque ela é um triunfo político. Valor de negociação internacional. Além do mais você hoje tem também um pouco mais de conhecimento técnico sobre a Amazônia.
[19] According to a Brazilian participant at the conference, this was probably due to the more co-operative attitude of the Brazilian government.
[20] The North-South divide has played an important part in debates on forests. Each group is heterogeneous; the 'North', for example includes both the USA and the Scandinavian countries, which hold very different views on both environmental and development issues. Nevertheless, each group is also characterised by some common concerns and opinions, which distinguish it in relation to the other group. When referring to the views of the North or the South this study refers to the views that characterised the position of the group as a whole.
[21] There were nine proposals for a global forest instrument, either a protocol or a convention: from the IPCC; the independent review of the TFAP; the WRI; the European Council; the G7 declaration in Houston – the same one that proposed the PPB; the European Council; the scientific and technical plenary of the Second World Climate Conference as well as the NGOs present at the conference; and the IUCN, International Union for the Conservation of Nature and Natural Resources (Humphreys, 1996).
[22] Apart from the Statement on Forests and Agenda 21, three other documents were signed at Rio: the Rio Declaration on Environment and Development, a set of 27

principles that should govern the environment and development of the world and that reflects the 'current consensus on values and priorities' in the two matters (Porras, 1993, p.21); the United Nations Framework Convention on Climate Change; and the Convention on Biological Diversity. The conservation of forests plays a role in all three documents, however, as the references to forests in these agreements were harmonised with the specific documents on forests, they are not to be reviewed in this book.

[23] New initiatives promoted by the programme are, for example, aquatic resources' management, sustainable forestry initiatives in the private sector, rainforest conservation corridors, community-level projects and environmental education.

3 The Rubber Tappers and the Development of Extractive Reserves

Introduction

In the 1980s, the rubber tappers and their leader Chico Mendes appeared in the media of the industrialised countries as the 'defenders of the forest'. They were seen as a traditional community who had been using their resources in common and living in harmony with nature until, encouraged by government policies, cattle ranchers arrived and occupied their lands. The ranchers engaged in the destruction of the Amazon Rainforest and the tappers defended it. The tappers' conservationist stance made possible the establishment of alliances between them and international environmental organisations, and in 1990, the state recognised their common property regimes through the creation of extractive reserves. There are now 37 extractive reserves, where over 50,000 commoners live.

This chapter examines the evolution of the rubber tappers' property rights institutions from the time of their arrival in Amazonia, in the late 19th century, until the establishment of the extractive reserves to understand how local and external factors interacted in the formation of a common property regime. I argue that the tappers' institutions have always been as influenced by external developments as by the characteristics of the tappers themselves and the features of their resources. The portrayal of the rubber tappers as defenders of the forest, for example, is a sign of how they articulated their struggle for secure rights with wider trends in society, such as the environmental concerns of the 1980s.

The analysis uses the theoretical framework outlined in Chapter 1. To recap: natural resources can be jointly used and conserved if either they are abundant or if co-users develop a robust common property regime. For resource users to develop such a regime, they must first perceive the need to do so, and the resource must have the potential to be managed as common property. If resource users are highly dependent on the resource for their survival, form a small and homogenous group, have a sound knowledge of their common resource, and have sufficient autonomy from the state to manage their resources, the potential for a regime to develop is higher than if these conditions are not met. The external context can also influence the development of these regimes by triggering changes in the circumstances and by influencing the capacity of resource users to deal with such changes. Hence, in examining the rubber tappers' struggle for secure land property rights, attention will be paid to the presence of external actors, government policies and, in more general terms, to the socio-political, institutional and legal setting in which the resource and resource users are embedded.

From private rubber estates to common property regimes

Since its origin in the late 19th century, the rubber trade has always been centred on vast rubber estates. Initially, these estates were privately owned by large landowners, the '*seringalistas*' or rubber barons, but over time, they became the common property of their inhabitants, the rubber tappers. This section identifies the factors that changed the property rights regime of the rubber estates.

Rubber estates may be defined as common-pool resources. They are large areas of tropical forest – between 130 and 700 sq. km – with a high concentration of rubber trees (Weinstein, 1983), which can be jointly used by several individuals as long as the activities practiced do not require the removal of the forest cover, which is the case with rubber tapping. Resource users must respect the maximum sustainable yield of the trees. If too much latex is extracted from rubber trees, they dry up; to avoid this happening the trees need to have 'rest' periods. The rubber tappers' work has remained the same since the time of the rubber boom: the tapper walks along the rubber trails (*estradas de seringa*) making cuts in the trees; he[1] then puts a small pot underneath the cuts and later collects the latex accumulated. Each tapper has a rubber stand (*colocação*), which is the basic unit of production of rubber tapping, and is formed by an average of two or three rubber trails some of them lying fallow; the allotment designated for agriculture; areas for fishing, hunting and gathering; the tapper's house and the site for the processing of rubber. The rubber tappers' holdings are on average 5.5 sq. km (Allegretti, 1989) and their shape is determined by the rubber trails. The trails of one *colocação* are generally intertwined with the trails of other rubber stands, thus the latter are better defined according to the trails, rather than to the land area they occupy. Because of the ecological interdependency of the forest, the intertwining of the rubber trails, and the distribution of water and other resources on the estate, the resource is not easily divisible.

Many rubber barons, specially the wealthier ones, delegated the management of the estate on a patron (*patrão*) and spent most of their time in Manaus, but others lived most of the year on their estates, where they had a house, a warehouse and other dependencies. *Seringalistas* and patrons controlled the tappers' individual use of the resource through a debt-peonage or semi-slavery system known as *aviamento*. Under *aviamento* there were no monetary transactions: the rubber tapper had credit to buy food and industrialised goods from the owner of the rubber estate, on security of the rubber extracted at the end of the harvest. The tapper could sell the rubber only to the patron who paid much less than the market price, and often discounted rent for the use of the land. Rubber tappers were generally indebted to the patrons and they could be arrested if they left the estate without paying off their debts. When the price of rubber was high, tappers were forbidden to practice any other activity for subsistence, and they were compelled to buy all agricultural goods from the patron. Threats of violence helped supporting the system.

The *aviamento* system, however, was also supported by social ties between the patron and the tappers. The patron, for example, was often the godfather of the rubber tappers' children (Hecht and Cockburn, 1989; de Paula, 1991; Allegretti, 1989). The patron also provided the necessary common services, such as clearing the forest paths, and in case of illness it was to the patron or his wife that the tappers would resort. The *aviamento* system was thus characterised by a combination of physical violence with social dependency, a fact that, as will be seen in later chapters, helps to explain some of the rubber tappers' attitudes with regard to the management of their resources.

By and large patrons were uninterested in the management of the forest (Almeida and Menezes, 1994). Nevertheless, there were some regulations on the use of the rubber trees, such as stipulations concerning the amount of rubber that could be extracted and the length of the fallow period for the trees. These rules were enforced by the patrons' employees. When the price of rubber was high, however, patrons sometimes encouraged the tappers to extract latex above the maximum sustainable yield, which resulted in the drying up of the rubber trees, a situation that was very frequent in the area that is now the Extractive Reserve Rio Ouro Preto (see Chapter 5). The criticisms made to private property in Chapter 1 can thus be observed in the context of the rubber estates. If the price of rubber was high, the most efficient solution for a *seringalista* could be to deplete his estate because the costs of overusing the rubber trees were lower than the benefits of selling large quantities of latex at a very good price. Hence, the reason why forests were conserved during the rubber boom was not that the rubber estates were privately owned and thus the barons carried most of the benefits and costs of overusing the resource. The main explanatory factors for the conservation of the Amazon forests during this period were, first, that the most profitable economic activity, rubber collection, did not require the removal of the forest cover. And, second, that given the abundance of available land, when the price of rubber was high a *seringalista* could choose between encouraging tappers to extract too much latex from the trees on his estate, or attracting workers to tap rubber outside his estate[2].

The development of common property arrangements on the Amazonian rubber estates was a gradual process that resulted from alterations in the local context of the tappers, which was in turn the outcome of changes taking place in other parts of the world. Although these changes impacted on all the rubber estates of Amazonia, common property arrangements only developed in some parts of the region. In more remote areas, *aviamento* still exists (Schwartzman, 1990; Allegretti, 1989), and whereas in western Acre, where the Alto Juruá Reserve is now located (see Chapter 5) rubber estates were privately owned until the mid-1980s (Feitosa, 1995), in eastern Acre informal common property arrangements began developing in the 1960s. The remainder of this section identifies the combination of factors that led to the development of common property regimes in the east of Acre, in particular in the Acre River Valley where the Extractive Reserve Chico Mendes was later established.

The process that led to the formation of common property institutions in Acre began with the end of the rubber boom in the 1920s, when rubber plantations were developed in Southeast Asia, ending the Amazon region's virtual monopoly. Rubber plantations – more profitable than collection of wild rubber – were tried in Amazonia but with no success. A plant fungus, *Microcyclus ulei*, attacks rubber trees in their natural habitat. Wild rubber trees are scattered among other trees in the forest, therefore the pest cannot destroy a whole Hevea population as it can do on a plantation where the trees are side by side (Dean, 1987; Hecht and Cockburn, 1989)[3]. The rubber trade regained some importance during World War II. The Asian supplies of rubber had been blocked by the Japanese, and the US made an agreement with the Brazilian government, 'The Washington Treaty', to reactivate rubber production in Amazonia. As part of this treaty, workers from other regions of the country were once again encouraged to go to the Amazon forests to work for the war effort as 'rubber soldiers' (*soldados da borracha*). With the end of the war the Amazonian rubber trade plunged again.

Because of the crises of the rubber business, many tappers left the forests, but those who stayed saw their autonomy increase. Many rubber barons and patrons abandoned the rubber estates and those who remained relaxed the strong labour control of *aviamento* (de Paula, 1991; Weinstein, 1983; Duarte, 1986). Rubber tappers were thus relatively free to develop their own property rights arrangements because they either had been left to their own devices or had sufficient autonomy from the patrons. During the period 1920 to 1940, rubber tappers diversified their economic activities, increased social ties among themselves, and established commercial bonds with other economic agents (de Paula, 1991). The patrons were having difficulties with the provision of goods, so they allowed and encouraged rubber tappers to practice other activities for subsistence. Besides, the tappers could no longer rely exclusively on the marketing of rubber for meeting their needs, and subsistence agriculture and collection of forest fruits was a way to supplement their income. The diversification of activities contributed to increase the tappers' autonomy because it helped them to break their dependency on the patrons' provision of goods. Moreover, it made them responsible for the overall management of the stands, and not just for the collection of a specified amount of rubber. As patrons had less power over them, the rubber tappers also started commercialising with middlemen, which further diminished the *aviamento* ties.

During World War II, patrons and *seringalistas* regained control of the rubber estates, but the level of domination they exerted over the tappers was less rigid than during the first few years of the 20th century, when the rubber trade was flourishing (de Paula, 1991). Whereas during the rubber boom, tappers were completely dependent on the patron because there was no law or independent entity that protected their interests, during WW II, labour relations between patrons and rubber tappers were officially regulated by a standard contract (*contrato padrão*) established through a government institution, the Rubber Credit Bank. Under this contract, the rubber tapper was a tenant who rented his stand from the patrons, and had the right to sell 60% of his collected *hevea* at a price equivalent to the price of rubber in Belém and Manaus. With regard to food provision, tappers were also more independent from the patron than before, since under the standard contract each tapper had the right to one hectare of land for agriculture, and was allowed to fish and hunt for commercial purposes (de Paula 1991; Hecht and Cockburn, 1989). It cannot be stated with certainty, however, to what extent the new labour relations actually followed the contracts; there is a shortage of documentation regarding the life of rubber tappers during this period. Nevertheless, de Paula (1991) has presented evidence of a number of conflicts between tappers and patrons, which indicates that when the contract rules were disregarded, tappers reacted against the patrons. Such conflicts suggest that the tappers were more autonomous in the 1940s than at the time of the rubber boom. Yet, whether their enhanced autonomy also led to collective action initiatives is not known.

With the end of World War II, as more *seringalistas* and tappers abandoned the rubber estates, *aviamento* ties got even looser. Rubber tappers could no longer be imprisoned because of debts and, in the 1950s the introduction of radio on the rubber estates broke, or at least diminished, their traditional isolation. Through the radio, the tappers could be informed of the rubber prices outside the estate, and so patrons and traders could not mislead them as easily as before (de Paula, 1991). Most conflicts between rubber tappers and *seringalistas* during this period were because of the price of rubber (Allegretti, 1979), and there are records of rubber tappers having positive financial balance with the patrons, thus being in credit rather than indebted to the patrons at the end of the harvest. These changes in the rubber estate context led to the development in the 1960s-70s of the 'autonomous rubber tapper'. 'Autonomous rubber tapper' is the term used to define the extractivist who does not depend on the patron and who is free to sell his product to other social agents, as well as to practice complementary activities such as collection of Brazil nuts (*Bertholletia excelsa*) and subsistence agriculture (Allegretti, 1989).

In the Acre River Valley by the late 1960s, there were large numbers of autonomous rubber tappers, who held their forests as common property. Their common-pool resources were either whole rubber estates or parts of large estates, which they jointly used under little or no control from a central agency. They used the resource according to a set of individual and common rights; the extraction of rubber was conducted on their own 'private' rubber stands while rivers and forest paths were used by all. Tappers followed the previously established rules concerning the use of rubber trees and respected each other's rubber trails, thus complying with the internal boundary rules of the estate. Common areas were used in a system that is better defined as 'restricted access' rather than as common property, since access to the common areas was restricted to the resource co-owners, but there were no regulations to harmonise the co-owners' use of such areas.

Several internal and external factors were crucial in the development of common property regimes on the rubber estates. The principal internal factors pertaining to the natural resource were the common-pool character of the rubber estates, which rendered their division in private plots difficult, and the existence of a plant fungus. This fungus was largely what prevented the development of rubber plantations in the region, which in turn was one of the reasons why the Amazon rubber trade collapsed. In terms of price, wild rubber could not compete with rubber from plantations. With regard to the resource users, the main change was that their access to information increased, partly because of external factors, such as the technological developments that facilitated the introduction of the radio on the rubber estates, but also because rubber tappers began socialising more, with fellow tappers and with other social agents. External factors were particularly important in the case of the rubber tappers because their economic activity was related to external markets. Although the tappers lived in considerable isolation, they never were self-sufficient communities and always maintained links to the external world. The rubber trade crises, which originated in the external context (the international market for rubber), were crucial to the rubber tappers because through a chain of events they increased their autonomy to manage the forest resources. Thanks to the difficulties of the rubber trade, the tappers who remained in the forest stop relying exclusively on the patrons for the provision of basic goods and for selling their rubber. Less dependent on the barons, tappers became responsible for a wider range of activities, and increased their level of contact with other agents, two developments that further enhanced their autonomy.

The tappers' common property regime of the 1970s, however, failed to meet many features of a robust regime. Their common rights to the abandoned rubber estates were not legally recognised. They lacked informal rules protecting the boundaries of their common resources and monitoring devices to secure their own compliance with the rules; besides, there is no evidence showing that the tappers had jointly decided to follow the existing rules. These rules were largely the same as those enforced by the rubber barons, and the tappers now followed them because they protected their trees and avoided conflicts with the neighbours. The absence of any initiative to develop a robust regime can be explained by the fact that, most probably, the tappers did not feel the need for one. Their resource was abundant: there was no competition among the tappers over its use and hardly any outsiders ventured into the area.

With regard to problems that tappers felt were important, such as their newly developed dependency on the middlemen, it was around the 1970s that the first initiatives were taken. Some extractivists, such as Chico Mendes, the rubber tappers' leader who later became known worldwide, were trying to organise their fellow tappers to set up co-operatives to commercialise their produce. They also tried to establish schools on the rubber estates – higher levels of literacy would enhance the tappers' capacity to assess their accounts with the middlemen (Mendes, 1989; de Paula, 1991).

Threat to the boundaries of the common-pool resource

In the late 1960s, in the context of the government policies for 'developing' the Amazon region, a highway was opened linking the capital of Acre, Rio Branco, to Assis Brazil – the BR 317 highway (see Map 5.1, page 161). Access to the rubber estates along this track was considerably facilitated and this, initially, further contributed to increase the autonomy of tappers in the area. They now had easier access to markets, and their dependency on middlemen was somewhat reduced. However, the increased autonomy of the tappers was soon overridden by other less positive developments. Easier access to the rubber estates through the BR 317 highway, and the new government policies, which encouraged investment in Amazonia, attracted many new comers to the region, who bought large tracks of forests along the newly opened highway[4]. As the boundaries of the rubber estates were unprotected, either legally or informally, the tappers saw their resources – and their livelihoods – threatened.

The threat to the boundaries of their resource was the main factor triggering the 20-year process that transformed the tappers' weak common property regime into a more robust one. The threat to their resource, however, did not result from one individual incident; it was triggered by developments in the national sphere and, in particular, by the new government policies. During this period, between the early 1970s and the mid 1980s, the state had a distinctly negative impact on the rubber tappers: it encouraged migration to their areas and failed to protect their scarce legal rights.

Property rights to the rubber estates were imprecise. Tappers living on the estates, like most peasants in Amazonia, were untitled occupiers of their plots (*posseiros*). They had usufruct rights to the area where they lived and worked, but the estates were formally public lands (*terras devolutas*) or belonged to the former rubber barons. The rubber barons were selling their estates cheaply because many of them had gone bankrupt. In addition to the already diminished world demand for Amazonian rubber, government help to the rubber sector was ending. In 1966, the Rubber Credit Bank had collapsed and, in 1967, the state monopoly of rubber terminated. The rights of the former rubber barons to the estates, however, were unclear. As we saw in Chapter 2, they often did not have legal titles to their estates, and when they did, the estates' area was not defined in terms of sq. km or hectares, which were the measures used to define landed property in the newly developed land market. Given the imprecise[5] definition of the area of their estates, land owners often claimed to own areas of forest that were larger than their rubber estates, a strategy known as the 'enlargement of lands' (Basilio, 1992). In the specific case of Acre, the situation regarding land ownership was further complicated because land titles had been granted by four different entities: the government of Bolivia, at the time that Acre belonged to Bolivia[6]; the independent State of Acre; the state of Amazonas, before Acre was made a federal state; and finally INCRA, the federal land agency. Sometimes, each of these entities had granted land titles to the same area of forest. Consequently, land titles in Acre amounted to more land than the area of the state (Basilio, 1992; ELI, 1994; Schmink, n.d.).

The new formal owners of the estates did not want extractivists[7] living in their lands. One of the defining features of common-pool resources is that they can be used in common; this, however, depends on the type of use of the resource. Rubber estates can be jointly used for extractivism[8], but joint use is impossible for simultaneously tapping rubber and rearing cattle: whereas the former requires the standing forest, the latter involves its removal. There were cases when the ranchers allowed extractivists to stay in some areas of their ranches (*fazendas*); there was still sufficient abundance of land for ranchers and extractivists not to get in each other's way. The presence of extractivists, however, interfered also with the market value of land. Land with *posseiros* was known to carry problems with it – either the new owner had to pay compensation for usufruct rights or expel the occupiers – and therefore had a lower price. The general policy of the ranchers soon became to expel rubber tappers from their lands.

Initially, many rubber tappers abandoned their stands to the new owners, and those who refused to leave did so individually. Resistance to eviction was not conducted through any form of institutionalised co-operation. The ranchers' strategy was to send their employees to the tappers' houses telling them to leave; if they objected the ranchers obstructed roads and forest paths, destroyed cultivation fields, and expelled the families with the help of gunmen (Duarte, 1986). Following the general pattern in the region, the expulsion of rubber tappers was extremely violent (Hecht and Cockburn, 1989; Schwartzman, 1992). Many extractivists left for rubber estates in Bolivia or migrated to nearby cities – between 1970 and 1980 the capital of the state of Acre, Rio Branco, doubled in size. The cities' infrastructure was insufficient to cope with such an increase in population and, moreover, rubber tappers lacked the necessary skills to find work outside the forest; consequently many went back to the rubber estates, establishing themselves in areas that had not yet been occupied by the ranchers (Hecht and Cockburn, 1989; Duarte, 1986; ELI, 1994).

Several factors explain the lack of collective action in the first instance of the cattle ranchers' invasion of the rubber estates. As mentioned earlier, the rubber tappers' common property regime was weak. It had not been designed with the specific purpose of protecting the common resource from outsiders, and therefore it lacked mechanisms to protect the estate from the entrance of outsiders. Moreover, regular contact among rubber tappers was infrequent, and co-operation among them was not institutionalised. Although tappers used the estate in common, they lived in relative isolation from each other (on average, there is one hour walk between rubber stands), and saw themselves as owners of their stands rather than of a common resource. In addition, as the tappers had never owned their stands (they had either worked as semi-slaves for the patrons or rented their stands from them), they were generally unaware of their usufruct rights as *posseiros*. Consequently, when the ranchers told the rubber tappers that they were the new owners of the estates, extractivists thought that ranchers had the legal right to expel them and that if they stayed on they would be infringing the law. Although rubber tappers were highly dependent on the resource and formed a homogenous group, without information about their rights, little contact among themselves, and having to face agents who were considerably more powerful than they were, their capacity to protect their resources was very low.

The development of collective action

Increased access to information and help from external agents enabled the tappers to engage in collective action and secure their permanence in the rubber stands. After trying to make a living in the cities, the tappers became aware that they hardly had any livelihood alternatives outside the rubber estates, and communicated this information to those who had remained in their rubber stands. Knowing how was life in the cities reinforced the extractivists' awareness of how dependent they were on the forest. Tappers also obtained access to information thought the action of external agents. The latter, moreover, encouraged the tappers to organise themselves and collectively manage their common resources, to improve their marketing of rubber, and to resist eviction.

Three groups of external actors were important for the rubber tappers: members of the Catholic Church, the Unions, and independent policy-makers.

In the early 1970s, under the 'liberation theology'[9], priests and nuns from various Catholic orders organised literacy courses on the rubber estates, and set up grassroots communities promoting religious teaching and social action (Mendes, 1989). These communities formed the basis for community development on the rubber estates; most leaders of the rubber tappers' movement were formed in Christian base assemblies, where they learned reading, writing, and the rudiments of community organisation. The Church was also the first to denounce the violence that the ranchers were using against the rubber tappers, helping them in this way to get their plight known outside the rubber estates. Most importantly, the religious informed the rubber tappers of their legal rights to the rubber stands (de Paula, 1991; Duarte, 1986). Knowing that they had legal rights to their stands bolstered the tappers' resistance. Whilst information on their legal rights did not help concerning forceful eviction by gunmen, it made it more difficult for the ranchers to lure the tappers into leaving their stands by telling them that they had no right to be there because they did not own the land. In the same way that information about the market price of rubber had helped to alter the tappers' business relation with the patrons, information on their legal rights contributed to change their approach to the cattle ranchers.

In the second half of the 1970s, CONTAG, National Confederation of Agricultural Workers, established rural workers' unions in Acre. The first union was established in Brasileia (see Map 5.1, page 161) in 1975 and two years later Chico Mendes helped to set up another one in Xapuri. The unions can only partially be treated as external agents. The decision to establish them on the rubber estates was taken outside the forest, but their establishment was conducted by rubber tappers' leaders such as Wilson Pinheiro, who was murdered in 1980, and the worldwide known Chico Mendes. The unions provided the organisational structure behind the rubber tappers' movement against land eviction. Hence, the structure of tappers' movement was exogenous, because it was adapted to rather than created on the rubber estates. On the other hand, the land issue was the motor behind the development of the rural workers' unions in Acre, whose members at the time were mostly rubber tappers. Prior to their establishment on the rubber estates, CONTAG unions were conservative, welfare-distribution agencies, which the tapers transformed into more radical organisations actively engaged in the fight for landed property rights (Hall, 1997b). Instead of being only supporting external agents, the unions were 'the main expression of the rubber tappers' resistance' (de Paula, 1991, p.83)[10].

The unions provided the organisational know-how for the rubber tappers, especially for those like Chico Mendes and Wilson Pinheiro who, already before the arrival of the cattle ranchers, were trying to get the tappers together to improve their social conditions (Mendes, 1989). The most important initiative in this field was Projeto Seringueiro (the Rubber Tapper Project), which began in 1980. The purpose of this project was to encourage tappers to set up co-operatives and its cardinal component was a literacy programme designed by the Ecumenical and Documentation Information Centre (CEDI), based on the teachings of Paulo Freire about empowerment through education[11]. Besides the Church and the Xapuri Union, the project counted with substantial co-operation from individual policy-makers, in particular anthropologist Mary Allegretti, who was later to set up a NGO, Instituto de Estudos Amazônicos (IEA), aimed at supporting the tappers. Apart from articulating literacy programs and encouraging the organisation of co-operatives, the establishment of unions was essential in leading the tappers' resistance against land eviction, including the staging of the *empates* or standoffs (Mendes, 1989).

Staging *empates* was one of the main strategies of the rubber tappers' resistance against eviction, and the first known instance of 'collective action' on the rubber estates. *Empates* are peaceful standoffs of extractivists against the clearing of the forest: rubber tappers, sometimes with their wives and children, go to an area of the forest that is about to be cleared and confront the workers, preventing them from cutting down the trees. The *empate*s were generally organised by members of the union, who when informed that a clearing was going to take place, would collect more details about the problem, and convene a meeting with the tappers. Afterwards, the leaders of the rubber tappers' communities and other union members would go from hut to hut telling the families about the day and place of the *empate*. During the standoffs, that in some cases lasted up to 3 days, and involved 60 or more rubber tappers and their families, the leaders tried to reach an agreement with the person in charge of the clearing. If no agreement was reached, the lawyers of the union took the case to court. The high level of participation recorded at the standoffs, the frequency of these initiatives, and their effectiveness indicate that free riding was not a problem. Free riding in relation to standoffs would have meant avoiding the risks involved in participating (participants in standoffs were sometimes taken to the police station and beaten up), but benefiting from the expulsion of the ranchers from the area.

The staging of *empates* represented an instance in which external actors supported the tappers and at the same time, the tappers adapted this external support to their own needs. Collective resistance may have been a concept that originated in the political teachings of the union. Given the characteristics of the tappers as a group – homogeneous, very aware of their dependence on the forest, with some (minor) experience in resistance – it is likely that even without the unions they would have eventually engaged in collective action. The unions, however, accelerated this process by providing them with information about how useful and effective collective action could be, and by teaching them the fundamentals of group mobilisation and organisation. The tappers, however, did not simply follow the 'instructions' of an external organisation. The type of action the unions suggested fitted with the isolated resistance the tappers were already carrying out before the establishment of rural unions in Acre. The rubber tappers adapted the know-how of the unions - collective action for welfare-distribution objectives - to their own needs, fighting for secure property rights to their forests.

Thanks to the *empates* and to the unions' lawyers, rubber tappers began obtaining compensation for having to leave their stands. Compensation was for the improvements made to the land (*benfeitorias*), such as the house, the animals, and the vegetable plot. Initially, compensations were pecuniary, and failed to solve the tappers' problems because with the indemnity received they could only migrate to a city, where, as shown before, it was virtually impossible for them to find a job. Rubber tappers were later compensated in plots of land, similar to those small farmers had – not delineated according to the rubber trails as the rubber stands were, and considerably smaller than the latter. After some time, many tappers sold their plots and moved away.

Agricultural plots were unsuccessful because they lacked the necessary conditions for the production and commercialisation of the produce (technical assistance, access to credit, roads for taking out the production, warehouses, minimum prices), and because they were based on the division of the area in private plots. Private property – often put forward as a solution for the management of common-pool resources – was in this case inadequate because if used for the extraction of forest products the resource was indivisible. The utilisation of the rubber estates, as we saw earlier, was based on a combination of individual and common rights. Whilst the institutional boundaries between the two were clear to the tappers, in terms of space there was considerable overlapping. Rubber trails of one stand, for example, crossed those of neighbouring stands; rivers that could be used by all cut across private stands. In the agricultural settlements, this intertwining was not possible, and as the areas granted to the tappers were smaller than their former rubber stands, vital resources were sometimes left out of their plots: the water stream was the private property of the neighbour. Private property also presented problems because the tappers found adaptation to this new form of social organisation difficult. The cultural knowledge needed in the agricultural settlements was very different from what they had learned from living in the forest (de Paula, 1991)[12].

The inadequacy of the agricultural settlements helped the tappers identify the features of their property rights institution. Autonomous tappers used the forest resources individually: each tapper worked for his own rubber stand and did not manage the common pool together with his fellow tappers. The rubber stands were to a considerable extent private property, in the sense that the family living in the stand carried most of the costs and benefits of using it. If the stand was cleared, it could not be sold for tapping rubber and, before the ranchers arrived, the only potential buyers were people interested in extracting rubber. If too much latex was taken from the trees, the stand would stop being productive, and moving to a new area required clearing a new set of rubber trails. The problems encountered in the agricultural settlements showed that although the tappers worked individually, their use of the forest involved more than only elements of private property. Given the indivisibility of the forest, the rubber tappers' property rights institution is formed by common rights to a large single area within which each family has rights to some rubber trails and a plot of land for agriculture. Everybody in the area has the right to use the forest paths and the fruit trees, although these common resources are often located in the area of the rubber trails, which are 'private' areas. A rubber tapper has thus individual rights to his rubber trails and plot for agriculture, but not to everything that is in the area of his stand. (When using the term 'rubber stand', we refer to the rights of the tappers to their trails, agricultural plots and other 'private' resources.)

The realisation that with financial compensation they could not make a living also helped the tappers define the features of their property rights institutions. In particular, it made clearer to them that the right to use the forest (usufruct rights) was more important than the right to commercialise the stands (private property rights). However, it is not clear whether all tappers thought that usufruct rights were more important than private property rights, or if only their leaders and some well-informed union members thought so. According to Rueda (1995), when in 1987 the first legislative attempt to recognise the tappers' common property rights was made, some tappers had wanted to have individual land plots. Rueda (1995) suggests that they may have been influenced by the government's policy of encouraging private settlements in Amazonia. Another possible explanation is that the tappers who showed preference for private property were not the same ones who had gone through the process of consecutively and unsuccessfully trying pecuniary compensation and land in agricultural settlements. In the Chico Mendes Reserve, the majority of the tappers interviewed considered that the right to sell land was superfluous, but that the commercialisation of stands was important – this distinction, however, will be further examined in Chapter 5.

To summarise, the principal factors that influenced the development of collective action among the rubber tappers were the policies of the national government, the legal setting, and the help tappers received from outsiders. Government policies such as opening roads across Amazonia and giving incentives for investment in non-sustainable activities encouraged the encroachment of outsiders in the tappers' forests, threatening their resource. In this respect, the case of the tappers is similar to that of many other commoners whose resources were destroyed because they became more accessible to outsiders, and the state (through the police and the tribunals) failed to protect their rights. It should be noted, however, that for the tappers, the opening of a road was also a positive factor because it diminished their isolation. Whereas many commoners are autonomous because they live in isolation, the tappers' autonomy was hindered by their isolation, which made them more vulnerable to other agents, such as patrons and middlemen. By contrast, contact with the external world, provided the tappers with information, which enhanced their capacity to negotiate with these actors. With regard to the legal setting, some components of it facilitated the struggle of the tappers, namely the recognition of usufruct rights. On the other hand, the lack of well-defined property rights contributed to the existence of a chaotic and violent land market in Amazonia, and to the destruction of the rubber estates. External actors' help was largely a positive factor. External actors facilitated access to information in a direct way, by for example informing tappers of their rights, and in an indirect way, by providing them with literacy courses, which enhanced the tappers' capacity to obtain and assess information. In this context, we should note that, as highlighted by Ostrom (1990), a very significant factor influencing the development of a more robust common property regime was knowledge of other alternatives (private plots and monetary compensation), which helped the tappers define their own property rights institution.

The wider political setting

In 1985, the rural workers' union from Xapuri decided to organise a national meeting of rubber tappers in Brasilia. In reviewing the reasons that led the unions to organise this event and the results of the meeting, this section attempts to show how the capacity of commoners to articulate their needs in the wider socio-political setting can be a determinant factor in the conservation of jointly used resources. The case of the tappers provides a good example of how commoners are affected not only by isolated external events, such as outsiders occupying their resource or external agents helping them. Commoners are also affected by wider trends in society. They need to devise their strategies within developments in the external context that are not always directly related with issues of common property such as the recognition of common property regimes by the legal system. In order to this, they need representation in the political arena as much as they need management systems for their resources.

Three factors triggered the organisation of the 1985 Meeting. First, the rural unions' realisation that collective action at the local level, namely the staging of *empates*, was insufficient for solving the tappers' landed property rights situation. Independently of the characteristics of the tappers as a group, of the features of their common resource, and of the help they received from external agents, the establishment of a robust institution also required state support to secure the resource boundaries and facilitate the tappers' joint use of the resource. Although the standoffs were achieving some objectives, such as avoiding the expulsion of the tappers from their rubber stands, they were failing to provide a long-standing solution (de Paula, 1991). Besides, the standoffs could not continue indefinitely because they posed an enormous strain on the unions' financial resources and on the tappers themselves. In the summer of 1985, two major standoffs had been organised, lasting approximately 2 weeks. During this time, the union had had to support the participants (e.g. providing them with meals during the standoffs). For the tappers, taking part in *empates* meant losing workdays during the period of rubber collection, which later resulted in additional difficulties for paying their debts with the middlemen. Rubber tappers' leaders also feared that with the frequency of standoffs, these would loose their impact. Conversely, a national meeting of tappers from all over the region could increase the lobbying capacities of the movement (de Paula, 1991).

Second, rubber tappers were concerned that they lacked a well-defined proposal to meet their needs. According to Chico Mendes:

> [a] moment arrived when we began to get worried, because we had got a fight on our hands, the struggle to resist deforestation, but at the same time we didn't really have an alternative project of our own to put forward for the development of the forest. We didn't have strong enough arguments to justify why we wanted to defend the forest (Mendes, 1989, p.37).

Through the unions and national organisations such as CONTAG, CUT (Unified Labour Centre, a confederation of unions), and the Workers Party (PT) the tappers had found a channel for their resistance, but they also needed a forum for addressing their specific demands, which were different from those of other rural workers (de Paula, 1991). A national meeting of rubber tappers could provide an arena to discuss and devise potential solutions for their own problems.

The national socio-political setting in 1985 provided the third motive for organising the National Meeting. With the end of military rule, government policies were changing and civil movements were organising to voice their demands, but the rubber tappers' concerns were largely ignored. For instance, although several rubber tappers attended the Fourth National Congress of Rural Workers, their specific demands were hardly addressed in the Congress' final resolutions. Also at this meeting, President Sarney presented the National Plan for Land Reform, an issue which at the time was quite central in the political agenda of the country, the tappers' landed property right requirements, however, were not mentioned. Likewise, in the draft of the new state policies for Amazonia, put forward by the government in 1985, no reference was made to the rubber tapper population (de Paula, 1991; Revkin, 1990). In the mid 1980s, the existence of the rubber tappers was hardly known outside some restricted circles of Acrean intellectuals and activists: '[...] for most of the authorities the Amazon region was just one big empty jungle. We wanted to show them that the Amazon was in fact inhabited – there were people living and working in the forest' (Mendes, 1989, p.38). Rubber tappers thus arranged that their national meeting should take place in Brasilia, to make their existence better known in the decision making centre of the country.

On the one hand, the tappers' initiative to organise a national meeting originated in the new possibilities offered by the altered national setting. In other words, the changes that had occurred in their external context made it possible for them to seek support at a national level, outside the local sphere where their struggle had so far developed. On the other hand, the objective of the meeting was to influence the external context and to induce the government to recognise their existence. The tappers therefore were not passive actors, who were merely influenced by developments in the external context. Their decisions were shaped by the wider national context as much as by internal factors, such as their characteristics as a group.

The National Meeting of Rubber Tappers took place in October 1985, in Brasilia, with the participation of 130 rubber tappers from several Amazonian states. External actors helped them organise the meeting. Rubber tappers received assistance from various organisations, such as CONTAG and the rural workers' unions of Xapuri, Brasileia and Assis Brasil. Through their contacts in the capital, namely anthropologist Mary Allegretti, the rubber tappers' Meeting had also the support of governmental agencies, such as the Pro-Memória Foundation, a branch of the Ministry of Culture, and of the University of Brasilia, which provided the place for the meeting. Funding came as well from international NGOs working with developing countries, including some that had previously supported the *Projeto Seringueiro*, such as OXFAM. As one of the main objectives of the Meeting was to influence the state, state officials were invited; the Rubber Tappers' Meeting was attended by governors of Amazonian states, senators, deputies, and government officials from the Ministries of Industry and Commerce, Education, Health, Agriculture, Agrarian Reform, and Culture (Mendes, 1989). In addition, there were several observers from abroad, especially members of international environmental NGOs.

The National Meeting of Rubber Tappers was successful in at least four ways. First, it served to define a solution to the tappers' problems: to create 'extractive reserves'. Second, it was at this meeting that a national organisation for representing the tappers' interests in the wider political setting was created, the National Council of Rubber Tappers (*Conselho Nacional de Seringueiros*, CNS). Third, the Meeting provided an opportunity for the tappers to articulate their specific demands within the national political context, and to make alliances with other actors, namely Indian groups and international environmental organisations. Fourth, the National Meeting contributed to make the tappers' struggle more widely known in the national and, especially, in the international arena.

The extractive reserve proposal

The central features of the proposal for extractive reserve were similar to the characteristics of the common property regime the tappers had before the arrival of the cattle ranchers. However, whereas that regime had not been designed by the tappers with the specific aim of securing the conservation of their forests, the explicit aim of the extractive reserves was to secure the conservation of the tappers' forests through an institutional arrangement devised by organisations of rubber tappers. As mentioned earlier, tappers defined the features of their arrangement by comparing the different solutions that outsiders had proposed. Through this comparison, they identified the features of their common property regime and the reasons why this regime was for them the best solution. The formulation of the extractive reserve proposal was a gradual process: the central features of the proposal were outlined in a number of meetings before and during the National Meeting[13].

The extractive reserve proposal was a demand for legal recognition and state support of a common property regime. The core features of the proposal were, first, that the extractivists' rights to the lands they had traditionally inhabited for generations should be recognised. The significance of this demand lies in the fact that the general policy for *posseiros* – untitled occupiers of their land – was to compensate them financially for the improvements made to the land or allocate them land elsewhere, sometimes in regions very different from their own. This, for example, was the case of landless farmers from the dry northeast of Brazil who in the 1970s were offered land in Amazonia. Second, the tappers' land tenure system should be maintained. The state should recognise their common rights to the common resource rather than grant private property rights to each rubber tapper family – this demand was as radical as the first one, for at the time, the concept of common property did not yet exist in Brazilian law.

The tappers participating at the Meeting were fully aware that the legal recognition of their landed property rights was essential for their livelihood, and that other development initiatives could only be carried out effectively once this issue had been dealt with. Yet, the other requirements of a robust institution, such as existence of conflict resolution arenas and support for internal monitoring, were not mentioned. The absence of these topics at the meeting, together with the tappers' history up to then, suggests that the potential destruction of the resource by the tappers themselves (the 'tragedy of the commons' scenario) had so far not been an issue on the rubber estates. For the tappers, other problems seemed more pressing. They demanded a rubber pricing policy, which should include support for the production and commercialisation of wild rubber; highlighted their lack of access to school and education services; and discussed the debt-slavery situation of many tappers, and that of the former rubber soldiers, who were not receiving their retirement pension. Another issue that was discussed was the model of development for Amazonia that had been followed during the last two decades and, in particular, the fact that this model had ignored the existence of tappers and other traditional populations in the region. What had started as a local movement for securing individual rights to rubber stands, had thus developed first into a struggle for recognition of what was a common property regime, and now into a set of demands which amounted to a requirement for sustainable development.

The National Council of Rubber Tappers

The National Council of Rubber Tappers was created to defend at the national level the specific interests of the rubber tappers as a separate category from other rural workers. The entities that had until then voiced the tappers' demands were the rural workers' unions, but whilst in Acre their members were mainly rubber tappers, in other Amazonian states most of their members were agricultural workers. At the national level, the tappers lacked political leverage within the unions because they represented a minority and agricultural workers sometimes ignored the extractivists' different needs. Another shortcoming of the unions was that they were not the most appropriate organisations for forging alliances with potential supporters within civil society, such as environmentalists groups (de Paula, 1991). There was a national organisation addressing the issue of rubber, the National Council of Rubber, but it catered only for the interests of the rubber estates' owners and the industrialists: when the rubber tappers sent a commission to discuss their demands the Council refused to receive them (Mendes, 1989; de Paula, 1991; Revkin, 1990).

The mandate of the National Council of Rubber Tappers included the promotion of health and education, and the establishment of co-operatives for the tappers. It also dealt with the different demands of the various rubber tappers' groups: the main problem for the tappers in Acre was the occupation of their lands by the cattle ranchers; in Rondônia it was the non-payment of the former rubber soldiers' pensions; and in other areas, like Alto Juruá, it was the continuation of debt-slavery (Revkin, 1990; de Paula, 1991). In all situations, however, the need for the establishment of extractive reserves was felt, and CNS became the main articulator of the campaign for the creation of such reserves.

Establishing alliances with landless farmers, Indian tribes, and environmentalists

From the onset, rubber tappers were conscious that they needed to present their demands as important not just for them, but also for society as a whole. Consequently, they began by basing their political strategy on the importance of rubber for the economy, but they soon realised that with the end of World War II rubber had lost its strategic importance[14]. The extractive reserve proposal was then articulated in relation to three other issues that in the mid-1980s had high political leverage: land reform, Indian tribes' rights to their lands, and environmental concerns.

The rubber tappers' demands included issues that were being discussed in the context of land reform, such as posseiros' rights to secure landed property rights. As in 1985 land reform was an important item in the political agenda of the country, demands for extractive reserves were initially presented as the 'land reform of the rubber tappers' (*a reforma agrária dos seringueiros*) (CNS, n.d.). The distinction between tappers and other workers was necessary because in general landless peasants wanted private instead of common rights to the land. An advantage of articulating the extractive reserve proposal with the debate about land reform was that the rural unions were familiar with this issue, and initially extractive reserves were enacted within the land policy of the country[15]. An important disadvantage, however, was that the dissimilarity between the tappers' demands and those of other landless peasants made lobbying for extractive reserves difficult.

The closest concept to the rubber tappers' demand was that of 'Indigenous Reserve', in which the land is legally granted to the resident population in function of their historical use of it. This was the only situation in which the state granted common rights to the lands where people lived instead of allocating private property plots (Allegretti, 1989; Revkin, 1990; Mendes, 1989). Articulating the extractive reserve proposal with Indian tribes' demands for land also made sense because although rubber tappers and Indians had traditionally been enemies[16], in the 1980s, both groups faced the same threats - clearing of their forests and expulsion from their commonly owned lands. Moreover, the fight of Indian Tribes for their lands had acquired political importance in the national context (Paula, 1991). This, in turn, was related to international interest on Indian tribes, which was closely linked to the concern with the conservation of the Amazon rainforest. In 1987, extractivists and Indians formed The Forest

Peoples' Alliance, and the demand for extractive reserves and indigenous reserves became part of the same strive for a system of development of Amazonia that would include the traditional inhabitants of the region.

The extractive reserve concept also included demands for the conservation of the forest. This made it possible to articulate their proposal with the environmental concerns that were gaining popularity in the national arena and, in particular, with the international campaign for the conservation of Amazonia. As we saw in the previous chapter, by 1985 the international campaign against deforestation in Amazonia was already under way: some months before the Tappers' Meeting took place, the environmental NGOs had succeeded in temporarily stopping World Bank funding for the Polonororeste. The initial contact between the tappers and the international environmental NGOs seems to have been made through personal contacts. Individual policy makers helping the rubber tappers, namely Mary Allegretti and Anthony Gross, were looking for potential donors for the Rubber Tappers' Meeting, and they sought the help of their contacts within the US environmental movement (Revkin, 1990). As the environmental NGOs' strategy involved finding partners in the countries where the MDB funded projects were located, for them, the tappers' Meeting signified an opportunity to find potential allies in Brazil.

Rubber tappers and environmentalists formed an alliance because some of their interests were similar – they were both trying to stop the development policies of the government, which were leading to the destruction of the Amazon forests. Moreover, given the political strategies of tappers and environmentalists, their mutual support was very useful for both. For the environmental movement the tappers were important because they helped them prove that, besides causing environmental destruction, the development policies supported by the MDBs had a deleterious effect on the local population. This was useful to respond to claims that the socio-economic benefits of the MDB-supported policies were more important than their environmental impact. Tappers also helped them show that the conservation of Amazonia was more than merely a concern of foreigners living in the industrialised countries of the North; local populations like them were also concerned with the destruction of the rainforest. Finally, the concept of extractive reserves was very attractive, for it seemed to provide an alternative model of development for the region: a model that considered the need for both economic development and forest conservation. The extractive reserve proposal appeared to epitomise the concept of sustainable development that was to gain so much leverage with the publication of the Brundtland report two years later (see Chapter 2). For the rubber tappers, the environmental NGOs represented powerful allies in their struggle for recognition and establishment of extractive reserves. They provided them with a certain degree financial backing, and made their cause known worldwide. Environmental NGOs had access to important political actors, such as decision-makers in the MDBs and senators in the US, who could in turn put pressure on the Brazilian government. They also had the means to lobby and experience in doing it. Hence, NGOs helped the tappers acquire political leverage by making their plight known and supported by public opinion in the industrialised countries, and by linking the tappers' plight to wider economic issues, including the renegotiation of Brazil's external debt.

The tappers' political strategy gradually became to focus on the importance of extractive reserves for the conservation of the Amazon, and internationally they became better known as a local community struggling to defend their forest, rather than as a group of landless workers fighting for land reform. This was not because their concern with the forest was the tappers' main defining feature. On the contrary: the rubber tappers' struggle had more in common with that of landless peasants in Brazil than with the environmentalists' campaign against deforestation. Like many landless farmers in Brazil, rubber tappers were fighting to have secure rights to the land where they worked; like them, rubber tappers were poor, had been marginalized by the political system, and were confronted with a high degree of violence. The ultimate objective of the rubber tappers and the landless farmers' struggles was to secure their livelihood. By contrast, what the rubber tappers and the environmentalists had in common was only that they both wanted to conserve the forest. Their wish to conserve the forest, however, was triggered by different motives. For most environmental organisations, the conservation of the rainforest mattered because of the global ecological importance of Amazonia, whereas for the rubber tappers the forest was important because it was their only source of livelihood. For the environmentalists, the conservation of the forest was a goal in itself, whereas for the tappers it was a means to an end; had they had alternative sources of livelihood it is not certain they would have wished to stay in the forest. For the environmental movement, Amazonia was one of many issues, but for the tappers it was their only concern. Alliances, however, do not arise only between social groups with similar interests; they develop among groups that can be useful to each other. The main reasons why tappers became known as forest defenders rather than landless workers was first, that the environmental movement had more political leverage than the land reform issue, and, second, that rubber tappers had more to contribute to the environmentalists' campaign than to the landless peasant cause.

Although the tappers adapted their political strategy to the wider context, and did so at the advice of external actors, they did not change their demands. The formal declaration of the National Meeting states that rubber tappers should be recognised as the 'true defenders of the forest', but also as producers of rubber (Platform of the CNS, 1985 in appendix E of Hecht and Cockburn, 1989). Besides, the manifesto includes all their original demands: that rubber stands had to be expropriated and demarcated according to the rubber trails, that their land should not be divided into colonists plots, and that areas occupied by tappers should be secured for the tappers' exclusive use.

The National Meeting represented a turning point in the tappers' struggle for recognition of their landed property rights. At the end of the meeting, the tappers had defined a solution to their problems and begun their campaign for the establishment of extractive reserves. They had also established alliances with other actors, in particular international environmental NGOs, whose help would prove crucial in triggering the establishment of extractive reserves. Moreover, they established a representative organisation at the national level, CNS, whose function is to articulate the tappers' demands within the government policies, the legal system, and the external context in general. Finally, the tappers' event in Brasilia helped to make their struggle more widely known at the national and, especially, at the international level.

The analysis of the National Meeting demonstrates the importance of the national socio-political context for the tappers. First, the political opening of the country was what made possible organising a national meeting. Second, the fact that rubber tappers set up an organisation to represent them at the national level suggests that they considered the national political setting crucial. Third, they articulated their demands with different issues, depending on their political leverage in the national sphere. In examining the 1985 Meeting, some interesting observations also arise regarding the role of the state in relation to commoners. One of the objectives of the meeting was to obtain state support, which suggests that co-owners of natural resources may need more than only autonomy from the state. Without state support, the tappers considered that they would be unable to secure their livelihoods, even if they could obtain legal back up for their rights. On the other hand, they were definite that they did not want the state to interfere with their property rights institutions. This confirms the argument presented in Chapter 1 that commoners may need state support to secure the sustainable use of their resources, but that the state should not take over the management of their resources.

It should also be noted that although the external context and external actors were important factors in the Meeting, the tappers were not passive agents in relation to either of them. On the contrary, the tappers' organisations made use of the external developments that were taking place at the time. The best example of this attitude was their decision to emphasise the ecological component of their proposal while at the same time maintaining their property rights demands.

The creation of extractive reserves

In 1987, INCRA answered the tappers' demands for extractive reserves by issuing an internal decree creating Extractivist Settlement Projects (*Projetos de Assentamento Extrativista*, PAEs) and, three years later, in 1990, President Sarney enacted a presidential decree creating Extractive Reserves. Whereas the PAEs were part of the land reform policy of the country, the Extractive Reserves were an instrument of conservation policy, and as the next chapter will show, they represented an important improvement in relation to the INCRA settlements. In reviewing the process that led to the creation of PAEs and Extractive Reserves, this section explains how socio-political and economic developments in Brazil and the industrialised world during the second half of the 1980s influenced the tappers' institutions. Four interrelated external factors are identified: the interest of the international community in the Amazon rainforest, the role of the media in publicising the tappers' plight, the MDB campaign, and the evolution of the relationship between the Brazilian government and the international community (see Chapter 2).

Soon after the National Meeting, and partly as a perverse result of President Sarney's proposals for land reform, violence in Amazonia increased. The ranchers were concerned that they could be more easily expropriated if their lands were occupied by extractivists, and redoubled their efforts to expel the tappers from the forest. In 1986, the rubber tappers organised several standoffs in which women and children also participated. This time, however, apart from confronting the ranchers' employees, the tappers' organisations also publicised these events and obtained media coverage of their efforts to protect the forest:

> At the same time as 100 or 200 colleagues are involved in the *empate,* standing in the way of the chainsaws and scythes, we aim to have a team whose job is to get information about what is happening back to Xapuri where another group will make sure it travels all over Brazil and the rest of the world. This is something we have only recently started to organise (Mendes, 1989, p.70).

The international environmental lobby and the foreign press were major assets for the tappers' fight and, according to Chico Mendes, the tappers received more support from abroad than from national entities (Mendes, 1989). The international media began referring to the rubber tappers and the extractive reserve proposals in 1986 (Allegretti, 1989). The National Meeting had made the tappers known outside Acre, and several individuals took an interest in the tappers' plight, helping to publicise their cause. Filmmaker Adrian Cowell, for example, who had been filming deforestation in Amazonia for several years, began in 1985 to film the tappers' struggle, focusing on the role of Chico Mendes, and encouraged officials from United Nations Environmental Programme (UNEP) to visit Xapuri (Revkin, 1990; Shoumatoff, 1991; Melone, 1993). Steven Schwartzman, from the Environmental Defence Fund, one of the leading NGOs in the MDB campaign, also took a strong interest in the tappers' struggle, and in 1987 invited Chico Mendes to travel to the US, where he attended the annual meeting of the Inter-American Development Bank. The Bank had decided to finance the highway BR-364 (see previous chapter), at the risk of creating in Acre the same level of destruction this road had led to in the neighbouring state of Rondônia. Mendes presented evidence of the negative impacts this road could have on the rubber tapper population, and asked for support for the establishment of extractive reserves; later in the year, he presented his case to the American Senate. In July 1987, Chico Mendes received two international awards: the Global 500 Award, given annually by UNEP to leaders in environmental action around the world, and the Better World Society Protection of the Environment medal, creation of US media tycoon Ted Turner. Mendes' visits to the US received considerable coverage from the national and especially from the international media, both in the US and Europe, where the 'rubber tappers' movement [was] now seen as a crucial defender of the environment in Amazonia' (Hall, 1997b, p.99).

It is a moot point whether INCRAs' enactment of the Extractivist Settlement Projects in the same year that Mendes travelled to the US and received two international awards were related or coincidental events. The discussions between INCRA officials and rubber tappers had begun before the tappers became internationally known, in 1985. Yet, the international attention the rubber tappers received in 1987 may have sped up the process of creation of the PAEs. Brazilian policy makers believed that without the political leverage environmental issues had acquired, the PAEs, which were the first item of land reform to include environmental considerations, would not have been accepted. The political leverage of the environment was in turn strongly related to the high profile this issue had acquired in the international arena.

In 1987, INCRA established two Extractivist Settlement Projects, São Luis do Remanso and Santa Quitéria. These PAEs, however, were established in areas of little conflict (Hecht and Cockburn, 1989), and violence in Acre continued, culminating in the murder of Chico Mendes in 1988. The conflicts that took place on rubber estate Cachoeira exemplify the problems faced by the tappers at the time. The inhabitants of this estate had agreed to refuse to sell their rubber stands to outsiders, but one of the group defected and sold his plot to a rancher, Darli Alves da Silva (Mendes, 1989). To stop the forest clearing that ensued, the tappers organised a standoff that triggered a high level of violence against them, to the extent that INCRA decided to expropriate the land, compensated Darci, and set up Cachoeira as a PAE. Most probably, tappers from Cachoeira lacked robust mechanisms to prevent defection, such as penalties for those who break established rules; but even if they had had such mechanisms, it is unlikely that they could have prevented Darli's occupation of the rubber estate. The violence used by the ranchers to expel the tappers was such – they hired gunmen to intimidate the rubber tappers – that even if none of the estate inhabitants had defected, the tappers would have been unable to protect *Seringal* Cachoeira without the support of the state and the police.

The PAEs represented a considerable victory for the rubber tappers' campaign, but they were only a temporary solution to their problems. The creation of extractive reserves addressed many of the PAEs' shortcomings. Upon the decree creating extractive reserves, President Sarney established four such reserves, including the Extractive Reserve Chico Mendes, and in the months before the Earth Summit, the new President Fernando Collor established another five extractive reserves. In the year of the Earth Summit, Collor also set up a state agency responsible for the administration of these environmental units, the National Centre for the Sustainable Development of Traditional Populations (CNPT). In the early 1990s, in the context of the Pilot Programme, Brazil and the G7 drafted a specific sub-project for supporting extractive reserves (see next chapter).

Two factors were decisive for the creation of Extractive Reserves in 1990: the high profile the tappers had in the international arena and the Brazilian government change of attitude towards the interest of the international community in the conservation of the rainforest. Although by no means the first rural worker to be murdered in land conflicts in Brazil, Chico Mendes' death in December 1988 created an international uproar and in the months that followed many journalists, environmentalists, and some senators from the American Congress visited Xapuri, the city were he had been assassinated. The murder of Chico Mendes received so much attention because by the end of the decade the rubber tappers' movement and Chico Mendes in particular were well known in the international arena. This in turn was related to the interest of the international community in the conservation of the Amazon rainforest. Hence, the different factors that triggered this interest, such as the MDB campaign, the publication of INPE's report, the droughts that occurred that year in the US, and the special place Amazonia holds in the mind of outsiders, enhanced the tappers' capacity to protect their resource because they made it possible for them to obtain international political leverage. Users of a common-pool resource which is not part of a globally important one, even if in a situation similar to the one of the tappers (i.e. all internal factors the same), would not have received the same level of international coverage and national political leverage. Consequently, their capacity to protect their resource would have been weaker than that of the rubber tappers.

By the end of the decade, the Brazilian government was gradually switching from a nationalistic attitude towards a more 'political' approach in which environmental concerns were no longer summarily dismissed as disguised attempts at the internationalisation of Amazonia. In this context, President Sarney took several initiatives to address environmental issues and one of those was to transfer the tappers' demands to the environmental policy of the country, legislating on extractive reserves. Given the high profile rubber tappers had acquired in the international arena, the creation of extractive reserves was an environmental measure with considerable international leverage for the government. A factor that contributed to the change of attitude of the Brazilian government was international pressure and the international economic context, which made the country more vulnerable to the international campaign against deforestation. The international developments reviewed in the previous chapter thus influenced the tappers directly, by making it possible for them to establish alliances with more powerful agents, and indirectly, by contributing to change the overall attitude of the Brazilian government towards the Amazon region.

The publicity the tappers received also helped them to get support from several international NGOs and Brazilian organisations[17]. With regard to the aid tappers receive from the Pilot Programme, this also occurred because the relationship between Brazil and the industrialised countries had improved and the tappers had a high profile in the international arena. Besides, SEMAM was politically powerful during the government of Collor (see previous chapter), and in anticipation of the Earth Summit the G7 countries and Brazil wished to show their commitment to the environment – it was during the months before the Earth Summit that the Extractive Reserves sub-project was most advanced.

To summarise, developments in the international context influenced the tappers' capacity to tackle their problems in the following ways. First, the strategy of international NGOs in the MDB campaign provided an opportunity for the rubber tappers to make alliances with powerful actors. Had the NGOs taken a more conservationist approach instead of advocating sustainable development it would have been difficult for the tappers to establish alliances with environmental groups. Second, the MDB campaign and the interest in forest conservation of public opinion in the industrialised countries contributed to change the national setting. Third, the change of attitude on the part of Brazil, which became more open to international demands to conserve Amazonia, and on the part of the industrialised countries, which realised that they had to financially contribute to the conservation of Amazonia, made possible for the tappers to obtain in addition to the protection of their property rights a considerable degree of support.

Analysis and conclusion

Upon the creation of a common property regime, resource users have to consider two different issues: how to harmonise their own use of the common-pool and how to protect their resource from outsiders. In reviewing the political process that led to the establishment of extractive reserves, this chapter has focused mainly on the first question, since during the period in question the tappers made no known attempts to articulate their joint use of the rubber estates. They held their resources in common and followed some rules that made their joint use of the forest possible, but these rules had not been created by them with the explicit objective of protecting their resource from their own overuse.

Chapter 1 argued that for a common property regime to develop it was necessary that the resource was amenable to be managed under common property. This was the case with the rubber estates, which are difficult to divide. With regard to exclusion of outsiders, this is relatively easy as long as the state recognises and protects the rubber tappers' property rights, since few outsiders would try to occupy lands that they cannot transform into pasture and sell in the land market. Moreover, if the entire estate is inhabited, tappers can easily monitor the resource borders.

Besides the features of the common-pool, the formation of common property regimes depends on the characteristics of the resource users and on the external context. With regard to the tappers' struggle to protect their resource against outsiders, many of the factors which the theory suggests are conducive to the formation of common property could be observed, such as perceiving the need to manage the resource, being highly dependent on the resource, and having access to information and to arenas for discussion. These factors were intrinsically related with developments in the external context. As predicted by the theory, the tappers only engaged in collective action once the need to do so was perceived, which was when the cattle ranchers occupied their lands. The ranchers' arrival, however, was an external factor triggered by developments at the national level, the new government policies of Brazil. The tappers' dependency on the resource was also a contextually dependent factor: if alternative sources of livelihood had existed in the cities, tappers would not have felt so dependent on the forest. Access to information was crucial in several occasions. Thanks to the introduction of the radio to the rubber estates in the 1950s, tappers became better informed about the market price of rubber and this in turn helped them obtain better prices for their produce; in the late 1970s, information about their legal rights to the rubber stands enhanced their capacity to resist eviction. Access to information was also contingent on external factors, such as external actors who facilitated information. Finally, with regard to arenas for discussion, their little access to it was related to the configuration of the rubber estates, which have no areas where tappers can regularly meet, like markets in agricultural villages. Arenas for discussion, such as those provided by the unions and the National Meeting, were crucial for the development of the reserves, and they were made possible because of both the efforts of the tappers and the changes in the external context, such as the political opening of the country, that facilitated that type of event.

In their fight to secure their resource against outsiders, the tappers' options were also influenced by the legal setting. The recognition of *posseiros*' rights to the land where they work in the Brazilian legislation provided the tappers with a means to resist eviction, whereas the absence of any legal item recognising common property rendered the task of legislating on extractive reserves difficult, as will be seen in the next chapter.

Overall trends in society were also very influential. For instance, the views of the state regarding the environment and traditional populations had a bearing on the tappers' management of their common-pool resources. In the 1960s and 1970s, the state approach to the development of Amazonia ignored the fragility of the region and the importance of securing the sustainable use of forests; traditional populations were at best ignored and normally considered to be 'backwards'. The state supported the ranchers because this fitted into their overall view of nature and society. Rather than being an isolated act in response to a specific struggle, the decision to establish extractive reserves was also embedded in the state's changed objectives. It was part of the government's new strategy regarding the environment and of the country's newly improved relationship with the international community. Likewise, the help of priests and nuns and of the Rural Workers' Unions was related to wider trends, such as the development of the liberation theology within the Catholic Church and the approach of CONTAG regarding the establishment of unions in places where there were disputes of workers.

The support tappers received from international NGOs and the attention the press gave them in the 1980s were also nested in a general trend taking place in society during that period, namely increased concern with environmental issues. In this context, events like the publication of the Brundtland Report and the Earth Summit, which fostered such interest in the environment, were crucial for the tappers. Equally important was the political leverage the Brundtland Report gave to the concept of sustainable development, since one of the reasons why extractive reserves received international support was because they were considered the epitome of sustainable development. The Earth Summit also had a number of implications for the tappers. The anticipation of the conference induced Brazil to set up extractive reserves and the G7 to provide support for the reserves in the form of the PP-G7. The debate on forests during the Earth Summit also had an indirect influence in the development of extractive reserves. The extractive reserve proposal fitted in with the arguments of both Brazil and the international community: on the one hand, the creation of reserves ensures the ownership of the forest for the nationals of the country and the Brazilian State[18]; on the other hand, it also fits with the argument that forests should be conserved for the benefit of humankind. The Forest Principles Agreement was a facilitative factor for the tappers because it provided them with a legal argument to justify their property rights.

The tappers were portrayed in the international media as defenders of the forest because this fitted with the trend in society at the time. This chapter has shown that the tappers' concern with deforestation was as important as their need for landed property rights, and that they wanted to maintain their traditional way of living largely because they lacked alternative livelihoods. They were portrayed as defenders of the forests rather than as fighters for landed property rights because this was a better means of obtaining support for their cause. Their environmental image was not, however, an imposition of outsiders who 'used them'. The tappers, or at least their leaders and advisors, knew that it was necessary to articulate their cause with those of society as a whole and, in the 1980s, the environmental issue was the one that had most political leverage.

A controversial issue in the debate about common property is whether for the development of such regimes resource users need to form part of small communities, whose members are relatively homogenous and expect to continue interacting over time (see Chapter 1). The evidence from the rubber estates suggests that regarding the protection of the resource against outsiders, access to information, awareness of danger, and external support are more important factors that the existence of small and homogenous communities. (Yet with regard to the harmonisation of their own use of the resource, as it will be discussed in Chapter 5, the existence of small and homogenous communities can be an important factor.) The process that led to the establishment of extractive reserves was not undertaken by a small community. Participants in *empates* formed small groups – 60 to 300 rubber tappers – but they did not all come from the same rubber estate. What made empates take place was the work of members of the union who went from stand to stand informing extractivists of the event. The help of external actors in terms of providing the tappers with an organisational structure, with information about collective action, and later with access to the media were the determining factors in the tappers' fight for protecting their common resource against outsiders. These factors, however, are alone insufficient to ensure the success of PAEs and Extractive Reserves – the next two chapters examine why is this so.

Notes

[1] Although some women also tap rubber, this activity is primarily performed by the male members of the household. See Campbell (1997) for an account of women's role on the rubber estates.

[2] Sometimes the surrounding area was owned by other barons, or occupied by Indian tribes powerful enough to keep the *seringalistas* at bay. In most cases, however, there was sufficient land for the barons to expand their territories.

[3] The issue of developing plantations in Amazonia has been researched by Dean (1987), as well as by other scholars such as Weinstein (1983). Although the main reason for the lack of plantations in the Amazon is the plant disease, other social and economic factors also played a role, e.g. the investment necessary for establishing plantations was large and it would thus take a considerably longer time to bring profit from a plantation than through collection of wild rubber (Weinstein, 1983, p.32).

[4] Apart from the federal incentives for buying land in the region, the state government of Acre also engaged in a massive campaign to attract southern investors (de Paula, 1991; Duarte, 1986). According to the Environmental Law Institute (1994, p.7), between 1970 and 1975 nearly 80% of the land of the state of Acre was sold to new private owners. By 1982, more than 100% of the state had been sold, with some municipalities (*municípios*) boasting that 160% or more of their land area had been claimed (Hecht and Cockburn, 1989).

[5] Imprecise in terms of the new measures used to define land; in earlier years, the definition of an estate according to the rivers and other landmarks it bordered had been sufficiently precise for the purposes of the rubber trade.

[6] The state of Acre belonged first to Bolivia. Later, Bolivia signed a contract with the Bolivian Syndicate – consortium with its headquarters in New York constituted by English and American capital – by which it rented Acre to the Syndicate. In 1902, Plácido de Castro promulgated the Independent State of Acre and finally, in 1903, Acre was annexed to Brazil.

[7] The term 'Extractivist' refers to those people whose activity is extractivism, collection of non-timber forest products. In this book, the term is used as a substitute for 'rubber tapper' that is in fact one category of extractivists only.
[8] Extractivism refers solely to the extraction of non-timber forest products, such as rubber, Brazil nuts, and other fruits, and does not include mining activities.
[9] A dissident strand within the Catholic Church that supports economic and political change in developing countries.
[10] A partir desse momento, a resistência dos seringueiros passará a ter como principal expressão as suas ações no movimento sindical (de Paula, 1991, p.83).
[11] This project also counted on modest funding from international organisations such as OXFAM and Christian Aid and subsequently from the federal government through the National Foundation of the Ministry of Culture (Mendes, 1989).
[12] 'Em primeiro lugar, ao sair dos seringais, esses trabalhadores, além de serem expropriados no plano material, objectivo – a perda das suas colocações – sofrem também uma expropriação subjectiva, que é a do seu saber, pois tudo o que haviam aprendido, através de experiências próprias ou herdadas de seus antepassados, para sobreviverem na floresta – tanto no dominio do trabalho até as relações sociais, culturais, etc, pouco ou quase nada lhes servia para enfrentar os desafios de uma forma de reprodução social tão diferente' (de Paula, 1991, p.129).
[13] At the National Congress of Rural Workers in May 1985, one of the propositions of the Acrean representative, a rubber tapper, was that the small plots allocated by INCRA did not work for the tappers; a special module of 3 to 5 sq. km should be considered for the extractivists (Allegretti, 1989). In the preparatory meetings for the National Meeting that took place in Rondônia, Acre and Amazonas there are also references to the basic characteristics of what would be later called extractive reserves. A rubber tapper from the union of Rondônia stated at the preparatory meeting that 'it is not a matter of owning the land, but of having the forest where the rubber estates and the rubber stands are, that the forest be demarcated as a forest reserve, so that the tappers can continue their extractivist activities'. A similar comment was recorded by Allegretti at the Xapuri preparatory meeting: '[it is necessary] to stop the clearing of the rubber estates. To expropriate taking into consideration the rubber stands'.
[14] Many tappers at the national meeting descended from those who had travel to Amazonia to be 'soldiers of rubber' and they still believed that their collection of rubber was of crucial importance for the country. Although the economic value of natural rubber had diminished considerably since the end of the war, they were strongly convinced that the importance of rubber for the national economy was their most important bargain point to obtain support from national entities. According to Allegretti (1989, p.21), 'when several speakers [at the meeting] presented evidence of the economic decline of natural rubber many rubber tappers, who had left the rubber estates for the first time in their lives did not accept this'. The same attitude on the part of the rubber tappers is described by Revkin (1990),

who states that tappers were urged by their advisors to switch their strategy away from rubber and towards their role as defenders of the environment.
[15] For more details on the legal aspects of this issue, see section 1 in Chapter 4.
[16] When in the late 19th century the future rubber tappers arrived in the Amazon forests, they often invaded Indian lands, and bloody encounters between the newly arrived tappers and the various Indian tribes were frequent.
[17] International organisations: Ford Foundation, MacArthur Foundation, Canadian International Development Agency, Environmental Defence Fund, Cultural Survival, Health Unlimited, WWF, UNEP, Survival International, Sierra Club; national organisations: Campinas University, National Bank of Economic and Social Development, state planning secretariats.
[18] Some national actors opposed to the tappers' demands, however, accused the extractivists of taking side with international actors trying to take over the Amazon.

4 Conserving the Forest in Extractive Reserves: An Assessment of Their Legislative Framework and External Support

Introduction

The outcome of the struggle for property rights of the rubber tappers from Western Acre was not only the recognition of their rights, but also the development of a legislative framework which facilitates the recognition of commoners' rights all over Brazil. Since the creation of extractivist settlement projects and extractive reserves, the common property rights of approximately 50,000 people have been legally recognised. Moreover, the extractive reserve model has attracted the attention of commoners and policy-makers outside Brazil, who have tried to replicate it in their own countries for it is believed that within extractive reserves local populations can both make a living and conserve the forest.

Previous studies on the capacity of extractive reserves to secure the conservation of the forest can be divided in two groups. One considers that extractive reserves cannot secure the long-term conservation of forests because tapping rubber, and collecting forest products in general, is economically unsustainable (Homma, 1989; Torres and Martine, 1991; Browder, 1992). Extractive products have a fixed supply, and if the commercialisation of the product is successful, the demand is likely to exceed the supply[1]. When this happens, the product starts being produced in plantations or an artificial substitute is found, and as the price of both the plantation product and the artificial substitute is generally lower than that of the 'wild' product, the demand for the latter diminishes. According to critics, wild rubber has been economically viable until now because of government subsidies; the price of natural rubber produced in the Southeast of Brazil is approximately three times lower than the price of Amazon rubber (Fearnside, 1989a). Critics of extractive reserves also argue that the environmental sustainability of extractivism is debatable, for although extractive activities do not require the removal of the forest cover (the main reason why they are considered environmentally sustainable), they can destroy the trees if too much of the product is extracted (Homma and Anderson, n. d.; Browder, 1992).

Supporters of the extractive reserve concept, including CNS, generally concur that in purely economic terms extractivism has serious shortcomings, but point out that besides the collection of forest products, there are other more profitable activities that can also be practiced in extractive reserves (FoE/GTA, 1994; Allegretti, 1994). With regard to the environmental sustainability of extractivism this will depend on whether resource users develop a robust common property regime.

The second group of studies has examined extractive reserves in relation to its socio-political aspects (Allegretti, 1989; 1994; Schwartzman, 1990; 1992; 1994). This line of analysis holds that by recognising the rights of extractivist populations, disputes over landed property rights and insecurity over land tenure, two important causes of deforestation, have stopped. Extractive reserves, moreover, form the basis for promoting sustainable development in the areas where they have been established because they are grounded on the proposals of grassroots organisations following their traditional land tenure systems. Yet, the sustainable use of jointly used resources depends not only on the existence of secure property rights. As Hall (1997b) has noted, the conservation of forests inside extractive reserves is also contingent on resource users overcoming the free-rider problem; the large size of most reserves, the relative heterogeneity of the reserves' inhabitants, and their isolation may render this task difficult; on the other hand, the rubber tappers' history of collective action may facilitate it. The overall success of extractive reserves – in economic, social and environmental terms –will depend moreover on whether they receive government support (Hall, 1997b).

Besides the size of the resource, the heterogeneity of its users, and their collective action experience, there are also other factors that influence the conservation of forests in extractive reserves. As argued in Chapter 1, the potential for a robust common property regime to develop depends as well on commoners perceiving the need to harmonise their use of the resource, on how dependent they are on the common pool, on their knowledge of it, and on their access to information. All these factors are in turn influenced by the economic and socio-political context, the legal setting, and the attitude of the state. This chapter assesses the extractive reserves' capacity to conserve the forests within their boundaries through an examination of the PAEs' and the reserves' legislative framework, and of the formal support they receive from the G7.

Extractivist settlement projects

With the enactment of the Extractivist Settlement Projects, the legal setting of the rubber tappers and of Brazilian commoners in general became significantly more facilitative than before – the PAEs were the first legal item in Brazil that recognised common rights to natural resources (Gomes and Filippe, 1994). In 1985, the concept of common property rights to land only existed in relation to Indian reserves, and landless peasants received land in settlement projects based on private property only (see previous chapter). As a result of the tappers' campaign for recognition of their rights, a working group formed by CNS, IEA, and technical staff from INCRA was set up to examine the issue and incorporate the tappers' proposal to the National Plan for Agrarian Reform. In July 1987, INCRA issued an internal decree (*portaria* n. 627 39th July 1987) constituting Extractivist Settlement Projects. Between 1987 and 1989, as can be observed in Table 4.1, ten extractivist settlement projects were established in the Amazon region.

Four items constitute the internal decree on PAEs. Item one states that the federal government, through INCRA, is the owner of the land, and that the extractive community has usufruct rights to it. This choice of state instead of communal or private ownership of the land was elicited by the characteristics of the tappers and of the *seringais* (internal factors), and by the changes that had occurred in the external context since the 1970s.

The rubber tappers' representatives and their advisers believed that state ownership would allow for more logistical support in terms of health, education, and production from the Brazilian State and from environmental organisations (Allegretti, 1994). State ownership also provided more land security than private ownership. As private ownership includes the right to sell the land, there was the risk that tappers could be pressured into selling their stands to cattle ranchers and land speculators, as it had occurred with many peasants in agricultural settlements (see Chapter 2). Besides, granting of private rights to the stands was difficult because the rubber estates are indivisible. Even if the private plots had been the same size as that of the original rubber stands (and not smaller, as were those in the INCRA agricultural settlements), the distribution of resources would have been disrespected, for common resources such as water cannot be divided between private plots. With regard to communal ownership, rubber tappers were not ready for it. Although the staging of standoffs was a proof that they could engage in collective action initiatives, tappers' leaders feared that once the threat of land expulsion was over, the sense of community that had developed among the extractivists would vanish; even in southern Acre, where tappers had most experience of group action, they were believed unprepared for communal ownership. In the 1970s, given the abundance of resources and the isolation of the rubber estates, a 'weak' common property regime had been sufficient to ensure the sustainable use of the forests. The situation in the 1990s, however, was different: outsiders such as loggers were interested in the rubber estates, and it was doubtful that tappers would be able to expediently develop a robust regime to deal with this and other new potential threats to their common resource.

Table 4.1 Extractivist Settlement Projects (PAEs)

Extractivist Settlement Project	State	Decree and date	Area (ha)	Population (number of families)
São Luis Remanso*	Acre	472/87	39 572	130
Cachoeira	Acre	158/89	24 973	80
Santa Quiteria	Acre	886/87	44 000	150
Porto Dias	Acre		22 145	83
Riozinho	Acre		35 896	120
Maracá I**	Amapá	1 440/88	75 000	214
Maracá II**	Amapá	1 441/88	22 500	94
Maracá III**	Amapá	1 442/88	226 000	760
Antimary	Amazonas	1 055/88	260 277	867
Terruã	Amazonas	1/89	139 235	426
Total			**889 598**	

Sources: Rueda, 1995; Allegretti, 1989.
* São Luis do Remanso was later incorporated in the Extractive Reserve Chico Mendes.
** The three Maracá PAEs later became the Extractive Reserve Rio Cajarí.

Items number two and three of the PAE internal decree define whom the co-owners are. Item two establishes that the use of the PAE area is transferred to a community by a contract between INCRA and the entity representing this community; item three states that participation in the community is limited to the traditional inhabitants of the area practising sustainable extractivism, and that the contract must say who the members of the extractivist community are. Including the name of the community members in the contract with INCRA was a way of protecting the traditional inhabitants from outsiders attempting to obtain rights to the common resource (ELI, 1994). Item four clearly acknowledges the existence of common property rights: it states that the community can use the total area and natural resources in it in a common property system, *condominio*.

The legislation does not mention the internal features of the common property institutions of the PAEs, such as rules to harmonise the use of the common resource by the co-owners themselves. As the tappers' regimes were weak, it can be assumed that with regard to internal features, the PAEs were weak institutions, too.

Although the PAEs decree represents an improvement in the legal setting of the rubber tappers, the Extractivist Settlement Projects in many cases unsuccessful: landed property rights' disputes and instances of resource depletion often continued in the areas where they had been established. The two main reasons for this were that the state failed to actively support the rubber tappers' rights inside the PAEs and that the extractivist communities lacked robust regimes.

The state agency INCRA contravened the principles of the PAE decree on several occasions (ELI, 1994; FoE/GTA, 1994; Rueda, 1995). For instance, instead of conceding rights of use to community organisations only, it also granted such rights to private individuals. In the extractivist settlement projects in Amapá, INCRA allowed the inclusion of migrants who had no experience in extractivism and no relationship with the existing community – hence disrupting the principle that only the established group of users should have rights of access to the resource. Because of this, the inhabitants of the PAEs in Amapá are still embroiled in land disputes with ranchers and loggers (FoE/GTA, 1994; Westermann, 1997). INCRA also supplied incentives to clear the forest and to develop unsustainable uses of the land inside PAEs, and rarely provided support for the establishment of PAEs; only the settlements in Acre received some support from INCRA (ELI, 1994; FoE/GTA, 1994; Rueda, 1995).

The problems of the PAEs confirm the argument made in Chapter 1 that the conservation of jointly used resources depends on both internal and external factors. It is a moot point whether the tappers, had they had robust regimes, would have been able to resist pressure from outsiders (supported by the state), and deal with new incentives for deforestation (also resulting from state policies). This is, nevertheless, a possibility. The literature on common property regimes provides many examples of commoners who were able to secure the conservation of their common resources in spite of new incentives for overusing it. The tappers in the PAEs had legal support for their rights, which at least in theory they could have used to oppose the entry of outsiders in their areas, even if the latter were backed by INCRA. They lacked, however, strong community organisations: in Amapá, communities began to form associations only recently, and in the Antimari and Terruã Projects – were extractivists experienced problems similar to those of Amapá – people also lacked any community organisation to administer the PAE (Westermann, 1997). Hence, either for lack of community organisation or powerless in relation to INCRA, the rubber tappers were unable to prevent the entry of new comers, which suggests that, as predicted by the tappers' leaders, extractivist populations lacked the organisational capacity to ensure the conservation of their resources in a changed environment.

Whilst the literature on common property institutions stresses the importance of autonomy, the tappers emphasise their need for external support. This is mainly because instead of state interference with their use of the forest, the tappers' main problem has always been their lack of access to medical and other social services. The tappers' experience has been of an absent state, which failed to protect their rights and left them subject to exploitation by patrons and middlemen. Nevertheless, the tappers did not demand state control of their resources, but state support only. The PAEs met only partially the tappers' requirements, since apart from state ownership of the land the decree lacks any other support measures. The changes that took place in the external context in the late 1980s and that led to the establishment of extractive reserves, made it possible for the tappers to obtain more support for their common property regimes, but they also appear to have restricted their autonomy to manage the rubber estates.

Legislation concerning extractive reserves

In reviewing the legislation on Extractive Reserves, this section shows first how developments in the legal setting that are not strictly related to recognition of common property can be useful for commoners, and then proceeds to examine whether the legislation on extractive reserves enhances the capacity of tappers to ensure the conservation of their resources.

One result of the changes that occurred in the socio-political context of Brazil in the late 1980s was the drafting of a new constitution (see Chapter 2), which provided an opportunity for strengthening the rubber tappers' common property institutions. The main constitutional basis for the creation of the reserves was Article 225. This article states that one of the measures the government must take to secure a healthy environment for the population is to define especially protected areas (Paragraph 1, III). Before 1988, conservation areas could be eliminated by decree, but with the new constitution environmentally protected areas can only be altered or eliminated if a law stating so is approved by the National Congress, which makes their existence more secure (ELI, 1994). By legally giving the extractive reserve concept the status of a conservation unit, the regularisation of the tappers' rights acquired a stronger legal base than the one they had in the context of the PAEs. The legal base of the latter was an internal decree of INCRA that could be revoked at any time by the director of the agency; the decree creating an extractive reserve can only be changed or revoked through a law approved by the National Congress. The decrees constituting each specific reserve cannot be easily altered either: they can only be changed if so stated by the National Congress, and even the Congress cannot alter the primary aim of the reserves which is to protect the sustainable livelihood of the inhabitants of the area (ELI, 1994).

The new constitution also served to attack the legality of cattle ranching, and obtain constitutional support for extractivism (ELI, 1994). Article 225 states that the Public Power has to protect fauna and flora by making illegal practices that threaten their ecological functions, destroy species, or imply cruelty towards animals (Paragraph 1, VII); hence, rearing cattle in Amazonia should be illegal because it threatens the ecological functions of the forest. Article 225 also declares that that the Amazon Rainforest is national patrimony, and requests all citizens to defend it (Paragraph 4), a statement that can be used to argue that extractivism should be constitutionally protected if it conserves the natural resource base of Amazonia. Finally, Article 225 (Paragraph 5) forbids the state to dispose of public land that is environmentally important in a way that would endanger the environment; hence, if rubber estates are environmentally important the state should not sell them to cattle ranchers because the latter threaten the conservation of the forest (ELI, 1994).

The first legal references to extractive reserves appeared in 1989[2], and on 30th January 1990, President Sarney signed the General Decree on Extractive Reserves (Decree n. 98.897/90). Upon the signing of the general decree on extractive reserves, president Sarney constituted the Extractive Reserve Alto Juruá, in the state of Acre, and in March four more reserves were established, including the Chico Mendes Reserve (see Table 4.2).

In addition to the federal reserves, the state of Rondônia also approved a law that authorises State Extractivist Settlements (*Assentamentos Extrativistas Estaduais*) and established five of them (ELI, 1994). In the

Table 4.2 Federal Extractive Reserves in Amazonia

Name	State	Date of Creation	Area (ha)	Population	Products collected
Alto Juruá	Acre	1990	506,186	4,170	Rubber
Chico Mendes	Acre	1990	970,570	6,028	Brazil nuts, copaiba, Rubber
Rio Cajarí	Amapá	1990	481,650	3,283	Brazil nuts, copaíba, Rubber, açaí
Rio Ouro Preto	Rondônia	1990	204,583	431	Brazil nuts, copaíba, Rubber
Pirajubae		1992	1,444	690	Cockle (seafood)
Ciriaco	Maranhão	1992	7,050	1,150	Babassu
Ext. Norte de Tocantins	Tocantins	1992	9,280	800	Babassu, fish
Quilombo do Freixal	Maranhão	1992	9,542	900	Babassu, fish
Mata Grande	Maranhão	1992	10,450	500	Babassu
Medio Juruá	Acre	1997	253,226	700	Rubber, fish
Tapajós-Arapiuns	Pará	1998	647,610	4,000	Rubber, fish, oils and resin
Lago do Cuniã	Rondônia	1999	52,065	400	Fish
Alto Tarauacá	Acre	2000	151,199	-	-

Sources: Ibama/CNPT 1994; Rueda 1995; Ibama website 2001.

months before the Earth Summit, among the several environmental initiatives taken by President Collor (see Chapter 2), were the constitution of four more federal extractive reserves and the establishment of an agency with the specific purpose of administering these reserves, CNPT (*Centro*

Nacional para o Desenvolvimento Sustentado das Populações Tradicionais).

Formally, CNPT is a 'facilitative' agency, which responds to the tappers' demands. Its function, as specified in the agency internal statutes, is not to manage the reserve on behalf of its inhabitants, but to provide support to the reserves' dwellers based on their own requirements. If, for example, rubber tappers propose a project CNPT officials evaluate it, harmonise it with available state funds, and monitor the use and consequent results of these funds. CNPT also promotes meetings between rubber tappers and relevant specialists in, for instance, agriculture extension, and organises management courses for the tappers (Rueda, 1995). In 1994, CNPT/Ibama enacted an internal decree outlining the formal procedures for establishing extractive reserves (*portaria* do Ibama n. 51 – 11/5/94).

The legal framework of Extractive Reserves comprises therefore two items: the General Decree on Extractive Reserves, which defines 'extractive reserves' and stipulates the requirements for any decree that enacts the establishment of a particular reserve, and Ibama's internal decree. Although these two items maintain the central features of the PAEs, there are four important differences between Extractive Reserves and Extractivist Settlement Projects:

- In ERs tappers' rights are more secure because the ER decree is more difficult to revoke than the PAE decree, and because the conservation of resources in extractive reserves is the concern of society as well as of the reserves' inhabitants.
- In ERs tappers' claims to land receive priority over competing claims.
- Rubber tappers can expect to receive more support from the state in the ERs than in the PAEs.
- The reserves' management system includes a set of rules and monitoring mechanisms specifically devised for the conservation of the jointly used common pool resource.

To what extent these differences between PAEs and ERs represent an improvement of the tappers' external context is discussed below.

The General Decree

There are several indications that in extractive reserves the state will promote the conservation of natural resources. The aim of extractive reserves is both to protect the rights of commoners and to conserve the environment. The General Decree stipulates that Extractive Reserves are areas designated for the sustainable use and conservation of renewable natural resources by extractivist populations (art. 1), and that they will be set up in areas where the sustainable use of natural resources is feasible, and which are of ecological and social interest (art. 2). Besides being a matter of concern for the reserves' inhabitants, the conservation of the tappers' forests is thus also a matter of importance for society as a whole, and so the state has a higher incentive to encourage the conservation of resources in reserves and comply with the reserves' regulations, created to ensure the ecological sustainability of the area. In addition, the state agency responsible for the reserves is Ibama (art. 3), whose primary function is environmental protection, and thus less likely than INCRA to provide incentives leading to resource depletion. Besides, INCRA had a poor record in relation to both PAEs and traditional settlement projects, and transferring responsibility for the tappers' commons to another agency was considered auspicious of better management of the rubber estates (ELI, 1994).

Commoners' claims to land are stronger in the ER framework because when expropriation is necessary, its justification is the protection of the environment for society (ELI, 1994). As by the end of the decade environmental protection had higher political leverage than land reform, extractivist populations' requests for recognition of their rights in the context of the extractive reserves' decree were likely to be more successful than the same demands in the context of the PAEs' internal decree. Furthermore, the extractive reserves' decree (art. 3) gives priority to the extractivists' common rights over those of other claimants. The state can only establish an extractivist settlement project after the land tenure situation in the area has been regularised. If there are conflicting claims to the land, the state must expropriate the land – a process that can be extremely lengthy – before formalising the extractivists' rights. The extractive reserve decree stipulates that the establishment of an extractive reserve should not require previous expropriation of the land; the argument for this is that in Amazonia most land titles lack legal basis (Allegretti, 1994; Menezes, 1994; see Chapter 2). Hence, the official document constituting a reserve must specify only which are the necessary measures that the state must take to establish a new reserve, measures that may include solving conflicts over land tittles in the area.

Besides the enhanced security of the tappers' rights in extractive reserves, two other components of the general decree indicate that within ERs rubber tappers will have strong incentives to ensure the conservation of their common pool resources: their rights are secure in the long term and they are conditional on the conservation of the resource. The state is the owner of the reserve; the entity representing the community has non-transferable usufruct rights over the area, for an indeterminate period of at least 60 years, by means of a contract with Ibama (*Contrato de Concessão Real de Uso*) (art. 4). This contract is based on the legal item 'concession of usufruct rights' (*Concessão de Direito Real de Uso*), which stipulates that the usufruct rights over private or state property are transferred to another entity for exploitation that is in the social interest. In this case, the 'social interest' is the conservation of the natural resource. The reserve inhabitants' property rights are thus conditional to the conservation of the resource. The contract between the community and Ibama includes an 'Utilisation Plan', which should be approved by the state agency, and that lists the rules that users should follow to secure the sustainable use of the natural resources base; if there is damage to the environment, the contract can be cancelled (art. 4). Ibama is responsible for monitoring compliance with the conditions specified in the contract (art. 5).

For common property institutions to be robust, they must be able to prevent overuse of the resource by both outsiders and co-owners even if circumstances change. The ER legislation largely provides for the first condition. Given how difficult it is to revoke the general decree and the decrees constituting each specific reserve, a decrease in the political leverage of the tappers or in international concern with environmental issues cannot alter the legal protection of the tappers' common pool resources. With regard to the second condition, preventing overuse of the resource by the tappers themselves, there are more provisions for addressing this problem in the ER decree than in the PAE's decree. In addition, the decree enacted by Ibama in 1994, includes several items aimed at encouraging the tappers to develop mechanisms for ensuring their own harmonious use of the forest.

Formal procedures for establishing extractive reserves

Ibama's internal decree, Manual for the Creation and Legalisation of Extractive Reserves, specifies in which cases the state should formalise a common property regime through the creation of an ER (Legal Creation); defines the procedures for implementing a co-management system between state and resource users (Implementation); and outlines the support that the state should give to the resource users (Consolidation).

Legal creation The legal creation of an extractive reserve consists in the publication of a presidential decree constituting the reserve. The process of creation of a reserve begins when a group of extractivists requests from

Ibama the regularisation of their lands in the form of an extractive reserve (Rueda, 1995). Hence, the state cannot create a reserve because Ibama or another state agency considers that a certain area ought to be protected; the creation of an ER is a response to users of a shared natural resource who require state back up for their property rights. CNPT is responsible for taking the final decision of creating or not creating an extractive reserve; the main criteria for deciding are the ecological value of the area and the existence of a common property regime, which also shows that Ibama does not intend to take control over the common resource. Extractivists must have an association that represents them, and should jointly manage the resource, or be willing to do so[3]. As the creation of a reserve does not require the existence of a regime that is robust, but only the willingness to jointly manage the resource, the ER legislation is an incentive for joint users of common pools to strengthen their property rights institutions.

Whether commoners whose natural resources are unimportant for society can also benefit from the extractive reserve legislation is not entirely clear. On the one hand, the General Decree states that reserves should be established in areas of 'social interest', which suggests that even if the resource is not nationally important, reserves may be established because support for a particular community is important. On the other hand, the fact that the ecological value of the area is one of the considerations for the establishment of a reserve suggests that if the resource is of value only for the direct users, it might be difficult for the latter to have their rights recognised. Twelve of the federal extractive reserves established so far are in the Amazon region, and the rest are in ecological important marine areas.

Implementation This consists in setting up a co-management contract between Ibama/CNPT and the organisation representing the reserve inhabitants. The stipulations for implementing an extractive reserve indicate strong support for the tappers' common rights. In principle, to conclude the contract between state and resource users it is necessary first to regularise the landed property rights in the area. For this, Ibama needs to examine the legal value of any landed titles to the area, determine whether or not to expropriate, and pay the necessary compensations to those with legal titles. The state then becomes the owner of the land in the reserve and can grant usufruct rights to the resource users. However, if there are conflicting claims to the land and the land regularisation process is likely to take several years, Ibama should proceed with the conclusion of the contract without waiting for the regularisation of the land rights (Ibama/CNPT, 1994), which indicates that the state gives priority to the informal rights of the commoners.

To conclude the contract it is necessary to address potential problems arising from the joint use of the resource by drafting an Utilisation Plan, which lists rules and monitoring mechanisms specifically aimed at securing the conservation of the common pool resource. The Utilisation Plan, which later becomes part of the contract, specifies thus the limits and conditions that apply to the extraction of forest products and the penalties that ought to be imposed in case of non-compliance. The Plan also states the conditions that apply to the transference of rubber stands and identifies the common areas of the reserve, which should be used in accordance with community rules. Co-owners can change the rules and the Plan must state the minimum number of inhabitants required to propose a change (Rueda, 1995). The basis for the rules listed in the Plan is the Brazilian environmental legislation, the environmental and socio-economic characteristics of the area (which should be described in the Plan), and the traditional use of the forest by the rubber tappers.

The Utilisation Plan includes also rules that apply to all reserves: the commercialisation of animals is forbidden, hunting with dogs is banned, fishing techniques must be sustainable, and mineral resources in the area cannot be exploited. There are as well some common prescriptions regarding monitoring, which correspond to the suggestions of the theory on common property. Compliance with the rules should be the responsibility of each inhabitant of the reserve, the entity representing the reserve inhabitants and, at a higher level, Ibama. If any other local organisation is responsible for sanctioning instances of non-compliance with the rules, the Utilisation Plan should state so. Sanctioning is gradual; if extractivists violate the established regulations, they will receive first an admonition, then a fine, next the temporary suspension of their rights, and finally their rights will be cancelled. The relationship between infraction and penalty is specific to each reserve and should also be stated in the Utilisation Plan. Regarding conflict resolution mechanisms, the CNPT decree stipulates that the organisation representing the reserve inhabitants should be the entity responsible for mediating in case of conflict.

Whereas in the General Decree on Extractive Reserves, there is only reference to the recognition of common property rights, the Ibama decree also acknowledges the tappers' individual rights to their stands. After concluding the contract between Ibama and the organisation representing the reserve, the latter can grant individual usufruct rights (*Títulos de Autorização de Uso*) to the inhabitants of the reserve. Individual usufruct rights are not transferable between living people; that is, tappers cannot sell their rights to the stands to other individuals. In case of death of the holder of the right to the stand, however, the right is transferred to the heirs – a clause that can act as an additional incentive to secure the conservation of the resource. In the absence of heirs willing to work in the forest, the rights revert to the state, which can give them to a third party.

Consolidation Under this item, Ibama's decree provides advice for improving the extractivists' social and economic conditions. The aim of these recommendations is clearly that the tappers themselves should manage the reserve; yet, some of Ibama's recommendations may hinder the tappers' autonomy. Ibama's decree suggests drafting a Development Plan for each reserve. The Development Plan covers six themes, four of which are directly related to the development of a robust regime:

- Measures to enhance the reserve inhabitants' capacity to manage the reserve (e.g. specify the type and level of training necessary).
- Measures to decentralise the social organization of the reserve in terms of area and activity to avoid the concentration of power in only a few.
- Regulations for managing the resource and ensuring community participation.
- Responsibilities of the state and other organisations involved in the co-management system.

The remaining two items of the Development Plan address the issues of 'production and commercialisation' and 'housing, transport, health and education', which also affect the robustness of the reserve albeit in an indirect way.

Is the legislation on extractive reserves more facilitative for tappers and commoners in general than the PAE decree? In many respects, the answer is yes. Commoners' rights receive stronger protection in ERs than in PAEs. There are also additional incentives for them conserving their resources, such as the conditionality of their rights on the conservation of the common pool. As it is an environmental agency, Ibama is less likely than INCRA to provide the reserves' inhabitants with incentives to practice non-sustainable activities or to encourage peasants without experience in extractivism to enter the reserves. Like INCRA, however, Ibama is also subject to the limitations of state agencies, such as lack of staff to monitor the use of large common pool resources, and insufficient knowledge of the resource and of how it is used (see Chapter 1). Another advantage of the legislation on reserves is that it addresses two issues ignored in the PAEs, namely that the conservation of jointly used resources requires the existence of rules and monitoring systems specifically aimed at ensuring sustainable resource use, and that to develop robust regimes commoners need support.

Yet, whether the Utilisation and Development Plans can facilitate or hinder the development of robust regimes is not clear. On the one hand, both Plans encourage the management of the reserves by their inhabitants. The Development Plan lists many activities aimed at encouraging commoners to manage their resources; training programmes, for example, promote contact among commoners and therefore enhance their capacity to develop a robust regime. On the other hand, the reserves' inhabitants seem to manage their resources only nominally. In theory, the rules of the Utilisation Plan are proposed by the reserves' inhabitants, which suggests that the basis for the Utilisation Plan are local conditions and traditional usages. But, given that in the past the rubber tappers had hardly managed their resources in common, the question arises as to what extent the rules and monitoring mechanisms of the Utilisation Plans are indeed those of the reserves' inhabitants. With regard to the Development Plan, a similar doubt arises: on the one hand, many items of this plan indicate that the state is aware of the tappers' lack of organisation, yet on the other hand the plan suggests drafting a management plan based on the tappers' proposals. If the rubber tappers' institutions are as weak as some articles of the Development Plan imply it is unlikely that extractivists will be able to propose a management plan for the reserve. We will return to these issues after reviewing the specific situation in the Chico Mendes Reserve.

Ibama's proposals for improving social services in the reserves and supporting production can also influence the robustness of the tappers' common property regimes. Poverty is a strong incentive for resource depletion. Even if tappers strengthen their common property regimes, faced with lack of access to medical services and poor return on their production, they will be inclined to improve their welfare by undertaking short-term strategies that may endanger the common pool. The conditions that apply to the joint use of a common pool depend on the use that is made of the resource; the tappers' regimes may be unable to prevent depletion of the common pool if they engage in other activities, such as logging, for which joint use rules are different. The price of rubber has been decreasing in the last few years, and given the new incentives faced by tappers, the decline of their economic situation has already given rise to instances of logging in the reserves. Commoners' economic development and their land tenure situation are interdependent: on the one hand, the tappers' economic development depends on the land tenure situation (without secure rights they cannot invest in the future); and on the other hand, the resilience of their land tenure system and the conservation of their resources depend on their economic development. Access to social services such as education, also influences the robustness of common property regimes because it facilitates access to information, which in turn enhances the capacity of commoners to strengthen their regimes and to tackle changes in the circumstances that threaten their resources. For all these matters, reserve inhabitants are thus likely to need outside support, which in the case of the Extractive Reserve Chico Mendes, it comes mainly from the Pilot Programme sub-project RESEX.

The Pilot Programme Sub-Project on Extractive Reserves (RESEX)

The importance of the PP-G7 lies in the direct impact it has on the extractive reserves and on the fact that it is a means by which developments in the external setting influence the rubber tappers. The establishment of the PP-G7 was due to the change of attitude of the international community towards Brazil's responsibility in the conservation of Amazonia, to the anticipation of the Earth Summit, and to Collor's strategy towards the country's international partners (see Chapter 2); these developments are thus also partly the reason why the rubber tappers receive support from the G7 countries. As the PP-G7 is the tappers' main source of external support, the developments in the national context that led to the PP-G7 are also external factors that indirectly influence the rubber tappers' common property regimes. The same argument applies to the factors that led to the inclusion of extractive reserves in the PP-G7.

Support for extractive reserves was already included in the first draft of the PP-G7, and in the years up to the Earth Summit, national and international actors concurred to rapidly advance this project. After the Earth Summit, RESEX was delayed, and it was only in 1995 that the first activities aimed at strengthening the rubber tappers' common property regimes were implemented. The main reasons for the inclusion, advancement, and delay of the RESEX subproject lie in the external context: developments taking place in the national and international arena during the first half of the 1990s have thus indirectly influenced the robustness of extractive reserves. The core reason for including reserves in the Pilot Programme was that national and especially international actors believed that tappers could contribute to the conservation of a globally important resource Amazonia. They thought so because 'firstly, the integrated processing of collected products appears to be well adapted to the ecological conditions prevalent in the highly diversified forest of Amazonia; secondly, organised groups of collectors are a viable form of protecting land from destructive use patterns' (WB/CEC 1991, p.31). Of these two motives, extractivism has been the main factor behind international support for the tappers[4]. Nevertheless, the change in people's opinion about common property also contributed to make it possible for the tappers to receive external support; if common property had still been considered as leading to the destruction of natural resources, extractive reserves could not have been logically supported.

Initially, support for extractive reserves was part of a project that also catered for national forests[5], but because of differences in the advancement of the two components extractive reserves became later a separate sub-project, called RESEX (Hagemann, 1994). The distribution of power within the Brazilian government in the early 1990s partly explains why extractive reserves received more support than other projects. Support for extractive reserves, as well as demarcation of indigenous territories and the creation of demonstration projects, was a priority for the Secretariat of the Environment (SEMAM), which in the early 1990s, during the mandate of President Collor, had considerable leverage within the government. Extractive reserves, however, were a politically difficult project, since other sectors of the government such as Itamaraty opposed the new model of development the reserves represented. Aware of this, SEMAM together with donors and the World Bank tried everything to advance the RESEX sub-project in the run up to the Earth Summit (Kolk, 1996; Hagemann, 1994).

Several factors contributed to delay the implementation of the RESEX sub-project until 1995[6]. One was that once the Earth Summit was over, interest in Amazonia and other environmental issues diminished considerably, in Brazil as well as in the rest of the world, and consequently interest in extractive reserves decreased (see Chapter 2). The change of government in Brazil (because of President Collor's impeachment) also contributed to reduce interest in the rubber tappers. The new Secretary of the Environment, Ambassador Perri, was not in favour of the Pilot Programme, which he thought put too much emphasis on extractive reserves, and government officials were worried that the RESEX sub-project would lead to the creation of additional reserves (Hagemann, 1994). It was thus only in May 1995 that the first disbursements towards the RESEX sub-project began. RESEX has four objectives:

> to test in four extractive reserves appropriate models of economic, social and environmental management; to improve methods and techniques used by traditional populations for the use of natural resources in tropical forests,

> through co-management between Government and society. The project aims at promoting the generation of rent, social equity and the spread of experiences concerning sustainable use of natural resources (GoB/BIRD/CUE, 1994:12[7]).

Chapter 1 argued that support for commoners should aim at restricting access to the resource by non-owners and at helping commoners strengthen their common property regime by providing them with information, arenas for conflict resolution, small-scale infrastructure, and support for establishing monitoring procedures. The RESEX sub-project undertakes all these tasks, but as shown in the next chapter it also interferes with the rubber tappers autonomy.

Sub-project RESEX is constituted by five components (see Table 4.3), and the first one, implementation of the reserve, aims at strengthening the reserve mechanisms to prevent overuse of the common pool resource by both outsiders and reserve inhabitants. RESEX provides support for the regularisation of landed property rights (e.g. it finances topological studies to demarcate the reserves' boundaries), and for strengthening the representative organisations of the reserves (e.g. it promotes meetings among the inhabitants of the reserve). The sub-project grants help also for the monitoring mechanisms of the reserve; it does so by strengthening the institutional capacity of the relevant state organisations and of the extractivist associations, and by training rubber tappers to become environmental monitors (*fiscais colaboradores*).

The support provided by the RESEX sub-project is based on the particular state of affairs of each reserve (GoB/BIRD/CUE, 1994; Irving and Millikan, 1997). Although the four reserves supported by the Pilot Programme were all created in 1990, they were at different stages of the implementation phase: in some, for example, the demarcation of the reserve boundaries and regularisation of landed property rights were incomplete; in others, such as in the Extractive Reserve Chico Mendes, there were still problems regarding the payment of compensation to former landowners. Concerning the entity that should represent the inhabitants of the reserves, this was not defined in all reserves; in the case of the Chico Mendes reserve, given its large size (it covers an area of nearly one million hectares) support was provided for the creation of three, instead of one, representative associations.

The second component of RESEX, community organisation, aims at improving the capacity of the reserve inhabitants to manage their resources, in order to both ensure the conservation of the common pool and improve the economic situation of the commoners. It does so by strengthening the

Table 4.3 Components and sub-components of Sub-Project RESEX

RESEX sub-project Components	RESEX sub-project Sub-components
Implementation of reserves	• Regularisation of land tenure • Strengthening of entities representing reserve inhabitants • Utilisation plan • Monitoring
Community organisation	• Physical and operational infrastructure of local associations • Training in administration, finance, accountancy and management • Education • Health
Improvement of productive activities	• Diffusion of improved technologies for subsistence and pilot activities • Improvement in processing and commercialisation of traditional products • Applied research on new potential economic activities • Improvements in storage, transport and communication
Management of natural resources	• Socio-environmental data base for reference for development plans • Research to support the production initiatives • Socio-environmental monitoring
Management and evaluation of sub-project RESEX	• Logistic support for CNPT • Training staff working in the project • Creation of information system for physical and financial data • Hiring of consultants for evaluating the project

Sources: GoB/BIRD/CUE, 1994; Irving and Millikan, 1997.

managerial capacities of the entities representing the reserves' inhabitants, of the co-operatives within the reserves, and of other organisations involved in the management of the reserve, such as Ibama, CNS, and the rural unions. In addition, it invests in physical infrastructure[8], sets up training schemes for the reserve inhabitants[9], and provides support for education and health.

Besides supporting the institutional aspects of the tappers' common property regimes, RESEX also addresses the commoners' economic needs: the third component of the sub-project aims at improving the return on the rubber tappers' productive activities. It supports extension work (e.g. cultivation of vegetable plots, agro forestry systems, rearing of small animals, and micro-fish farms), invests in the amelioration of rubber and Brazil-nuts processing techniques, and facilitates commercialisation of the produce. RESEX also supports applied research on economic activities that could improve the tappers' return on their work without damaging the forest. Alone, the collection of rubber is unprofitable, and tappers cannot make a living on it, yet when practiced in combination with other activities such as agriculture, small-scale farming, and extraction of other forest products, rubber tapping can be economically viable[10] (Schwartzman, 1994).

The fourth component of the sub-project RESEX is environmental management, which includes the development of a referential database to evaluate the evolution of the project, the implementation of a global system of environmental monitoring in each reserve, and a plan for the sustainable development of the reserve. This plan's objective is to examine what can be accomplished concerning the sustainable utilisation of natural resources in the area; for example, what are the potential impacts of the new production and transport developments in the area, how to minimise these impacts, which activities are most appropriate for the area. Under CNPT supervision, outside specialists together with people living in the reserve and representatives of the reserve inhabitants carry out the necessary studies for drafting the plan.

Broadly speaking, the RESEX sub-project was a positive development in the external context of the rubber tappers. It contributes to enhance the tappers' exclusion rights and includes items specifically aimed at encouraging them to strengthen their common property regimes. Moreover, much of the aid provided deals with issues that the tappers could not have tackled alone because of lack the financial and technical means, such as development of alternative economic activities, provision of detailed information on the environmental possibilities of the area, and supply of infrastructure. RESEX, however, presents also some important problems, which will become apparent when examining the specific features of the Chico Mendes Reserve in the next chapter.

Conclusion

In examining the legislative framework of extractive reserves and the support they receive from the Pilot Programme, this chapter has first shown how changes in the wider national legal setting and developments in the national and international socio-political context contributed to shape the formal institutional arrangements which regulate the use common pool resources in extractive reserves.

The new 1988 Constitution facilitated the tappers' capacity to obtain the recognition of their rights by giving a stronger legal basis to environmental units, by making the Amazon rainforest national patrimony, and by acknowledging that resource use can also contribute to natural resource conservation, as long as the use is environmentally sustainable. These items also helped to define the parameters of extractive reserves, since rubber tappers and their advisors used what was available in the constitution to develop the legislation on extractive reserves; for instance, they made ERS environmental units because the latter had a stronger legal back up than land reform items. Not only the legal setting influenced the rubber tappers, but the rubber tappers also influenced the legal setting. The tappers helped to change the national legal framework and in so doing altered the range of choices of commoners all over Brazil. The concept of common property of natural resources was recognised for the first time in the PAE decree, a legal item specifically designed to answer the demands of the rubber tappers, but which also facilitates recognition of common property rights of other users of common pool resources. Likewise, the extractive reserve legislation is useful to other commoners besides the rubber tappers, such as fishermen and collectors of forest fruits.

After the establishment of the extractive reserves, the socio-political context continued to influence the capacity of rubber tappers to manage their resources. The balance of power within the Brazilian government in the early 1990s was determinant for the inclusion of extractive reserves in the Pilot Programme, and later, after the Earth Summit, changes in the original composition of the government were largely responsible for the delay in the implementation of the RESEX sub-project. The priority that environmental issues received in relation to other matters in the international arena also explains the support that tappers receive from the G7. In the run up to the Earth Summit, the environment was one of the G7 priorities, and the Pilot Programme was established; after the Earth Summit Eastern Europe had more priority than the environment and, according to some observers, this influenced the disbursements for the Pilot Programme.

Second, this chapter has provided a preliminary assessment of whether the legal framework and aid that the reserves receive facilitates the conservation of natural resources. Chapter 1 argued that a facilitative framework should secure the boundaries of common pool resources against non-owners, and help commoners harmonise their use of the resource, but leave them sufficient autonomy to manage their resources.

The legislative framework on extractive reserves and the aid provided by the Pilot Programme go a long way to protect the reserves' resources from non-owners. Extractivists' rights receive priority over other claims, both upon the creation of the reserve and upon the establishment of the contract between Ibama and the reserve inhabitants. Property rights are granted for the long term and once granted they cannot be easily removed. There is a governmental body specifically aimed at supporting the reserves' inhabitants and that it is part of an environmental agency, which suggest that the state will comply with the legislation and grant property rights only to the members of the established community. The legalisation of the tappers' rights also diminishes the pressure from outsiders, since those interested in cattle ranching and in land speculation have little incentive to illegally occupy the reserves. The problem that remains is that of loggers, and this is something that Ibama alone cannot deal with it; the reserve inhabitants need also to monitor the boundaries of the common pool. The PP-G7, however, besides supporting the regularisation of the tappers' rights, also trains them to monitor their common resources.

When the first reserves were established, the rubber tappers' common property institutions were not sufficiently robust to deal with potential threats to their natural resources; in several ways, both the extractive reserves' legislation and the Pilot Programme address this problem. First, as the state is the owner of the reserves, the tappers cannot defect by selling their plots to outsiders, which was the only serious defection problem identified at the time. Second, the legislative framework conditions the tappers' property rights to their sustainable use of the resource base: if they destroy the resource, they loose their rights. Third, there are rules for using the resource and monitoring mechanisms to ensure compliance with such rules. Fourth, both Ibama's internal decree and the RESEX sub-project include a number of items specifically designed to encourage the tappers to strengthen their common property regimes.

In spite of all the provisions made by the Extractive Reserves' legislation and the RESEX sub-project, the conservation of the resource will depend in the last instance on how robust the extractivists' common property regimes are. To assess this it is necessary to examine not only the formal framework of the reserve, but also what rubber tappers actually do. In practice, do reserve inhabitants monitor the boundaries of their common pool or is Ibama that does it? With regard to the joint use of the forest by the extractivists, there are also some questions: if tappers did not have robust mechanisms for harmonising their use of the common pool before the establishment of the reserves, to what extent are the rules and monitoring mechanisms of the Utilisation Plans indeed those of the tappers? Are the stipulations of the legislation and RESEX sub-project appropriate to the specific features of each reserve? Is decentralisation of power, for example, necessary in all the reserves? The effectiveness of the activities designed to encourage reserve inhabitants to strengthen their regimes is likewise doubtful. According to the literature on common property, the development of a robust common property regime depends on the autonomy that external actors leave to the resource users. Although the aim of both the extractive reserves' legislation and the RESEX sub-project is that tappers should manage their resources, whether tappers have indeed sufficient autonomy to do so is unclear. Do tappers have autonomy at present or is autonomy only a future aim? If the tappers are not autonomous, should they have more autonomy, as the theory on common property regimes would suggest? That is, can they ensure the conservation of the rubber estates if outsiders refrain from attempting to strengthen their supposedly weak common property institutions? The next chapter will examine the actual characteristics of the Extractive Reserve Chico Mendes, and in so doing it will attempt to answers these questions and to re-assess the external context of the reserves' inhabitants.

Notes

[1] Homma refers to the supply of extractivist products from the natural forests; as the natural forest cannot 'increase', the supply of extractivist products presents little elasticity to changes in demand.

[2] The enactment of the new constitution made it necessary to introduce some alterations to the Law of the Environmental National Policy (LPNMA – *Lei de Política Nacional do Meio Ambiente*). The first reference to extractive reserves in the federal legislation is found in the alterations made to LPNMA in 1989 (law number 7.804 of July 1989). Article 9 determines that one of the instruments of the National Policy for the Environment is the creation of conservation areas and extractive reserves: 'Article 9: the instruments of the National Policy of the Environment are ... VI the creation of territorial spaces specially protected by the federal, state or municipal public power, of relevant ecological interest and extractive reserves' (Gomes and Felippe, 1994:76). However, it was only in September 1989, that a working group formed by state officials from Ibama and INCRA, CNS representatives and members of Mary Allegretti's NGO, IEA, began preparing the decree that should constitute the new legal item.

---

[3] Extractivists' applications to have their land made into an extractive reserve must include a statement that they have common property regime or that they are willing to do so. Ibama must confirm and complement the users' statement by visiting the area and gathering information concerning the level of organisation of the community and how the inhabitants use the natural resource base.
[4] With regard to biodiversity conservation, which is one of the reasons for international interest in the maintenance of the forest, extractive reserves are not necessarily the best solution. Because of the nature of extractivism, ERs are established in areas of low biodiversity, since from an economic perspective the best areas are those where there is a high concentration of single species (Browder, 1992). On the other hand, however, biodiversity does not require only the conservation of areas that are particularly rich in species, but the conservation of the forest as a whole. In this respect as well as in relation to the role of the forest in the global climate, extractivism can play a significant role (de Almeida, 1994).
[5] National Forests are conservation units where productive activities are also allowed; the difference between National Forests and Extractive Reserves, however, is that the latter are based on granting rights to the population already living in the area, whereas in National Forests anyone, or any organisation, can undertake the sustainable exploitation of the forest (Allegretti, 1994).
[6] In early November 1992, according to an internal survey of the Ministry of the Environment, there was no progress concerning the Extractive Reserves sub-project (Hagemann, 1994). In October 1993, a new version of the sub-project RESEX was presented to the World Bank. This version incorporated the results of the pre-investment phase as well as the comments made by the World Bank on the previous proposal, and counted with the contribution of CNS (GoB/BIRD/CUE, 1994). In April 1994, the World Bank appraised the latest draft of sub-project RESEX and declared it effective in February 1995.
[7] O objectivo geral do Subprojecto Reservas Extractivistas é testar, em quatro RESEX, modelos apropriados de gerenciamento econômico, social e ambiental aperfeiçoando os métodos e procedimentos utilizados pelas populações tradicionais, na administração dos recursos naturais renováveis nas florestas tropicais, através da co-gestão entre Governo e sociedade. Pretende-se viabilizar geração de renda, equidade social e difusão de experiências, tendo como base a conservação dos recursos naturais (GoB/BIRD/CUE, 1994, p.12).
[8] E.g. constructing central offices, facilitating transport and bettering communication services.
[9] In community organisation, administration, organisation of the production, commercialisation, financial administration and design of projects.
[10] According to Torres and Martine (1991) an economy based on the extraction of a combination of wild products, poly-extractivism, can be environmentally and economically sustainable. Demand for each product ought to be too small for their domestication to be worthwhile, since otherwise artificial producers would take over the collectors' market; demand for all the products together, however, can be

sufficiently large to make their trade profitable, or can at least guarantee a decent income for the collector.

5 A Common Property Institution in Practice: The Extractive Reserve Chico Mendes

Introduction

Established in 1990, the creation of the Extractive Reserve Chico Mendes (ERCM) regularised the landed property rights of the rubber estates of Southern Acre, where the rubber tappers' struggle against the cattle ranchers had started, and of many other *seringais* in the vicinity. With the establishment of the reserve, the quality of life of rubber tappers in the area has undoubtedly improved: they have secure property rights to their stands, and for the first time in their lives they receive external aid, mainly through the PP-G7. The conservation of the forest also seems more assured than before the establishment of the reserve, since tapping rubber does not require the removal of the forest cover, and ranchers and land speculators have been evicted from the area. In sum, given the provisions of the legislation and the help provided by the PP-G7, designed both to protect the reserve from outsiders, and to encourage rubber tappers to harmoniously explore the forest, natural resources in the ERCM, as well as in all other extractive reserves, ought to be sustainably used.

Yet, there are reserves' inhabitants who have sold timber to loggers and others who clear relatively large areas for agriculture. In the Chico Mendes Reserve in particular, an important problem is that many rubber tappers are abandoning the reserve and trying again to make a living in the cities; since 1995, nearly 400 families have left the reserve. This chapter examines the actual as opposed to the formal institutions of the Chico Mendes Reserve, and explains how an institution that appears sustainable on paper, can in practice have important shortcomings. The chapter assesses the capacity of the ERCM to ensure the conservation of the forest in the long-term, and identifies some of the failings that the extractive reserves' legislation and RESEX have – this provides an insight into the relationships between external and internal factors in the context of 'established' common property regimes.

I argue that the ERCM is a weak common property institution, and that this is largely because the rules and monitoring mechanisms of the formal institutional framework were not developed by the reserve's inhabitants. Therefore, some of these mechanisms are inappropriate to local conditions, and, besides, rubber tappers consider them as external rules for which they are not responsible. I also show that although the objective of the legislation and the PP-G7 is that tappers should manage their reserve, they leave them insufficient autonomy to do so, but then, I question the real importance of autonomy: can the tappers ensure the conservation of the reserve by themselves?

The chapter is structured in three parts. The first examines the jointness and exclusion conditions of the Chico Mendes Reserve, the characteristics of its population, and the co-management system of the reserve. The second part reviews the incipient and informal institutions within the many rubber estates that form the ERCM, and the role that external factors have played in their development. The concluding part presents a brief overview of other extractive reserves.

The bulk of the analysis of the ERCM is based on interviews carried out by the author with approximately 100 forest dwellers[1] and 10 key interviewees[2].

PART I – THE EXTRACTIVE RESERVE

The resource users and the common-pool

With an area of 9,705 sq. km (Feitosa, 1995), and inhabited by 1,465 families (CNPT/Ibama, c2000), the Extractive Reserve Chico Mendes belongs to the category of large common-pool resources shared by many users. Consequently, it is relatively difficult to manage: monitoring mechanisms for large common-pools tend to be more complex than those for small common-pools; and when co-owners are numerous they need to spend more time designing monitoring mechanisms than when they form small groups and can rely on social pressure alone to ensure compliance with the rules.

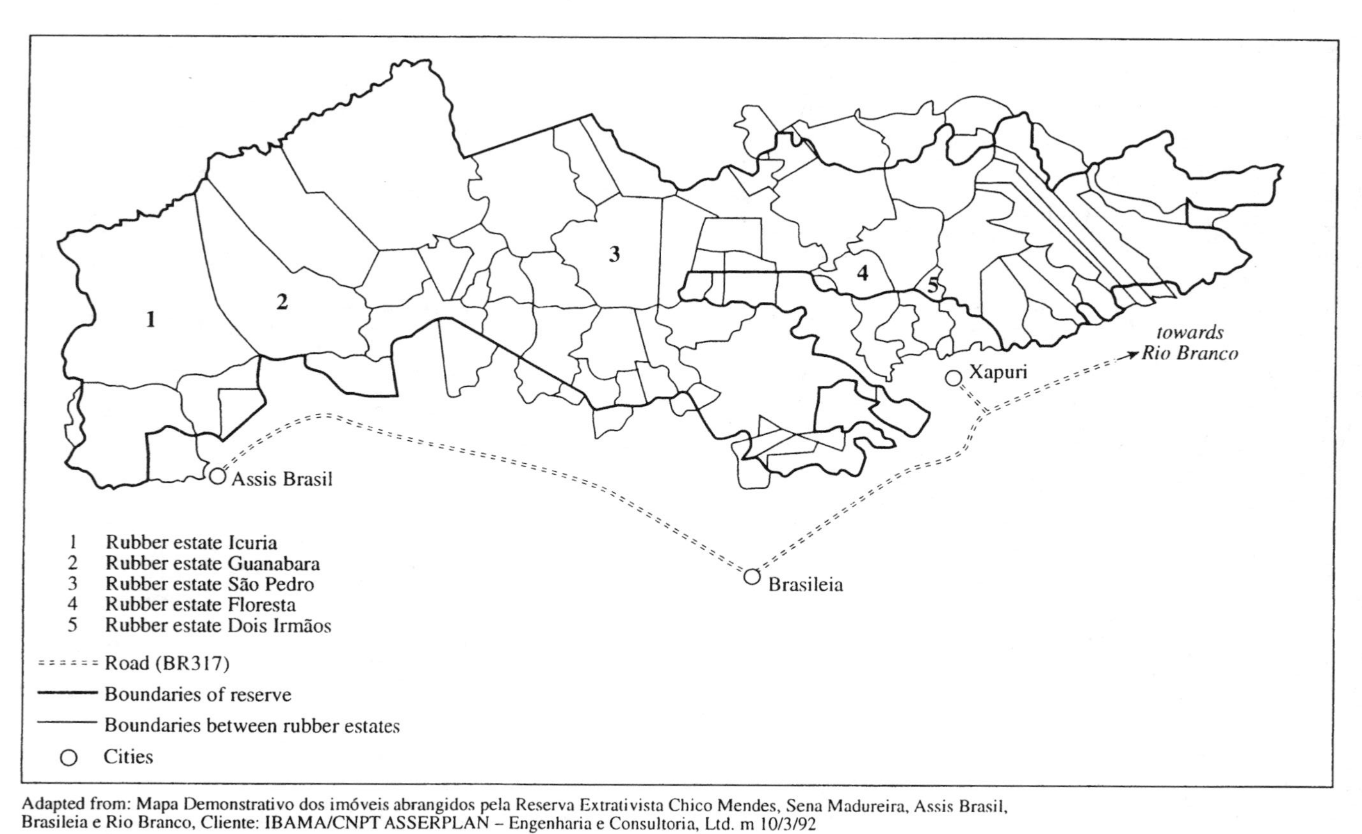

Adapted from: Mapa Demonstrativo dos imóveis abrangidos pela Reserva Extrativista Chico Mendes, Sena Madureira, Assis Brasil, Brasileia e Rio Branco, Cliente: IBAMA/CNPT ASSERPLAN – Engenharia e Consultoria, Ltd. m 10/3/92

Map 5.1 The Extractive Reserve Chico Mendes: boundaries of the reserve and of the rubber estates and former ranches

The approach of the rubber tappers' organisations upon the creation of the reserve explains many of the characteristics of the ERCM and its inhabitants. These organisations developed out of the landed property rights conflicts that took place inside what is now the reserve, on the rubber estates near the road BR 364 in the municipalities of Xapuri and Brasileia (see map 5.1). They fought, however, not only for recognition of their own members' property rights. When the government decided to create extractive reserves, their objective was to include as many extractivist areas as possible in what would become an extractive reserve. They knew that the problem they had faced with the arrival of the cattle ranchers would be repeated in any extractivist areas left outside legal protection, and if areas bordering their own stands were occupied by ranchers this would put pressure on their forests. Besides, including the largest possible number of extractivists areas in the future reserve was in itself an objective of the tappers' organisations, whose purpose from 1985 onwards was to improve the socio-economic conditions of the extractivist population in general through the establishment of extractive reserves.

One effect of the rubber tappers' organisations' approach is that the inhabitants of the ERCM form a large group with little experience of collective action. Less than a third of the tappers living in 1995 in the ERCM participated in the standoffs, and of these only a very small fraction followed the political and bureaucratic process that culminated in the establishment of the reserve. Most rubber tappers know that standoffs against ranchers took place in the reserve, and that it was because of them that they have secure rights to their rubber stands, which they highly value, but only a minority of the tappers endured the perils of collective struggle. The inhabitants of the ERCM, however, also exhibit a characteristic that is conducive to the development of robust common property institutions: they form a relatively homogenous group, with similar concerns, since the majority commercialises rubber and Brazil nuts, have an agricultural plot, hunt for consumption purposes, and share a 'common memory' of successful resistance against cattle ranchers (Feitosa, 1995).

Another outcome of the approach taken by CNS and the unions concerns the reserve boundaries. The boundaries of the ERCM correspond neither to natural boundaries – the areas inside and around the reserve present the same ecological features – nor to boundaries established by an informal institution, for no informal institution has ever had jurisdiction over the nearly 10,000 sq. km that form the reserve. When Ibama created the ERCM, the area that is now the reserve was a conglomeration of approximately 50 former rubber estates and *fazendas* (see Map 6.1), inhabited by autonomous and quasi-autonomous rubber tappers. The largest organisation of which they were members was a rural union, which represented at most a third of the area that is now the reserve, and generally less than that, since only some of the tappers were union members. The area of the reserve does not correspond fully to the extractivist areas either: whilst deforested areas have been included in the reserve, some rubber estates with tappers living on them had been left out, an error that occurred because Ibama had based the drawing of the reserve boundaries on aerial photographs without *in situ* confirmation (ELI, 1994)[3].

Both the characteristics of the resource users and of the reserve boundaries influence in turn the exclusion conditions of the ERCM. Excluding outsiders is difficult because of the absence of clear natural boundaries, the density of the forest, and the large size of the reserve. As the landscape is similar on both sides of the reserve limits, it is necessary to physically demarcate the boundaries. Monitoring access to the reserve is also difficult because forests are so dense that intruders can enter unnoticed at almost any point along the reserve boundaries, and as the boundaries are so extensive, it would be very costly for an external agency to monitor them. For the reserve inhabitants it is easier to monitor entry of outsiders; those tapers who live near the reserve boundaries can notice outsiders in their daily walks along the rubber paths. Even so, outsiders can enter through non-uninhabited areas; poverty can increase this problem because it is the main incentive for tappers abandoning the forests, and as more areas are abandoned, it becomes more difficult for the remaining tappers to prevent entrance of outsiders.

The joint use of the forest in the ERCM does not mean that that all tappers extract rubber from the same trees. The forest is 'divided' in rubber stands, and each family has individual rights to their stand; they can only extract rubber from the trees in their stand, which no one else has the right to use. Tappers cannot, however, act independently, since as an ecological system the forest is indivisible: the use each tapper makes of his stand has implications for the stands of all other tappers and hence on the reserve as a whole. Besides, rubber stands do not form clearly delimited wholes; they are intertwined. Moreover, tappers can use resources from each other's stands, such as fruit trees. The 'division' of the reserve in rubber stands is thus better understood as a specific allocation of rights based on the characteristics of the forests and on the use the tappers made of it. In the same way that villagers who jointly use a well may stipulate that each family has the right to use the well only once a day and take at most three buckets of water, rubber tappers jointly using the forest have the right to extract rubber only from their own stand and cannot clear the trees of their stands. This allocation of rights, however, was not entirely decided by the rubber tappers; *seringais* have been constituted in rubber stands since the time of the rubber barons. When in 1985 the rubber tappers' organisations devised the extractive reserve model, they decided to maintain the system of the seringais for conservation and other purposes, but only a minority of the reserve inhabitants participated in this decision. Hence, most rubber tappers interviewed were mainly concerned with their own stands (individual resource) and demonstrated little interest in the management of the forest (the common resource). In general, rubber tappers are individualistic (Hall, 1997b); for instance, outside the close family (parents and unmarried children), they rarely share rubber trails and agricultural plots[4].

The rubber tappers' economic activities allow for the sustainable joint use of the forest as long as clearings for agriculture remain small and that the maximum sustainable yields of the rubber trees and the game are respected. Even in the absence of a robust common property regime, tappers are unlikely to exploit the resource to its limit as in the case of the herdsmen described by Hardin (1968); if a family exhausts their rubber trees, they will have to clear a new set of rubber trails in a different area, which involves a considerable amount of work. Besides, the utilisation of the rubber stands is curbed by the number of family members able to work. Sometimes, stand-owners hire young tappers to work in their rubber trails, but employees are in short supply because given the low capital investment which is necessary to 'own' a rubber stand, young men move to their own stand as soon as they get married.

The tappers' joint use of the forest may, however, become unsustainable, if they switch to economic activities with a lower jointness potential, such as small-scale cattle ranching and logging. The inhabitants of the Chico Mendes Reserve are very poor, which is an incentive to engage in activities with a higher return even if only in the short term. The price of rubber has been decreasing for many years; the middlemen buy the rubber at even lower prices, and charge inflated prices for commercial goods. Furthermore, tappers have little access to social services – in 1994,

93% of the population was illiterate, and in 1995, only 43% had access to medical services in the city of Xapuri (ELI, 1994, p.55). If a change in the circumstances provides the reserve inhabitants with the possibility of increasing their earnings through non-sustainable activities, many will abandon extractivism thinking that once the resource is exhausted they can move to a city, where they expect to have better access to medical services and to schools for their children. Apart from being a crucial end in itself, addressing the issue of poverty in the reserve is thus also necessary for securing the sustainable use of the forests in the reserve.

The review of the characteristics of the reserve and its inhabitants suggest that the ERCM meets most conditions that make co-management an appropriate regime for the conservation of a jointly used resource (see Chapter 1). First, the common-pool resource is extremely large: a small group of tappers may set up an arrangement to secure the conservation of their own area of the forest, but unless the remaining inhabitants of the reserve also use their resources sustainably, their stands will risk depletion. It is, however, difficult (but not impossible) for small groups of resource users to establish such mechanisms for an area beyond their direct control, which is why in such cases external help is welcome. Second, as a group, the reserve inhabitants present characteristics that render difficult the development of a robust common property regime. The reserve inhabitants tend to focus on managing their own stands rather than the common areas. They form a large group spread over an extensive area and, given that the means of transport is walking, contact between tappers from different areas is rare. Even among tappers living nearby, contact is relatively infrequent since it does not occur via their work, which is individual, and neighbours live at least one-hour walk away. Third, the conservation of the forests is influenced by factors that are largely beyond the means of a local institution to address, such as access to health and education, and the world price of rubber. A system of co-management appears thus adequate to the ERCM, but whether the specific system in place in the reserve is conducive to the sustainable use of the forests is, however, a different matter.

Co-management in the Chico Mendes Reserve

This section examines the formal institutional arrangements of the ERCM in light of the information gathered through the interviews with the reserve inhabitants, and concludes that the forest dwellers participate little in the management of the reserve, which indicates that the co-management regime of the ERCM is weak.

The formal arrangements of extractive reserves include the following items (see Chapter 4):

- Delimitation and protection of boundaries.
- Participation of reserve inhabitants in the co-management system through their representative organisations, and
- Drafting of an Utilisation Plan where the rules, and the monitoring, and conflict-resolution mechanisms of the reserve should be specified.

Boundaries

Well-defined and protected boundaries are important for two interrelated reasons: they are necessary to ensure that the resource is used only by a restricted group of users, and they are important for co-owners to know what they are managing, and for whom they are managing it. By 1995, at the time of the fieldwork, the boundaries of the reserve were formally well defined, and regularisation of the tappers' exclusive rights was virtually complete (see Table 5.1)[5]. Yet apart from those living on its borders, the

Table 5.1 Implementation of ERCM - Chronology

Dates	Events
March 1990	Decree establishing the Extractive Reserve Chico Mendes
September 1991	Demarcation of reserve boundaries begins
November 1991	Conclusion of reserve boundaries demarcation
February 1992	Information on land titles completed
1992	Completion of reserve population studies
End 1992	Expropriation actions of private lands inside the reserve begins
November 1994	Creation of Associations of Inhabitants of the Reserve Utilisation Plan discussed with reserve inhabitants
April 1995	Utilisation Plan approved by Ibama
April 1995	Implementation of PP-G7 sub-project RESEX begins

Sources: ELI, 1994; Feitosa, 1995; GoB/BIRD/CUE, 1994.

large majority of the interviewees did not know the boundaries of the reserve. One reason for this may be the inadequate description of the boundaries in the Utilisation Plan: the limits of the reserve are described in cartographic terms – difficult for a lay person to understand, let alone a mostly illiterate population – instead of terms of reference familiar to the forest dwellers, such as 'the reserve ends in rubber stand so and so, in seringal x' (the names of the rubber stands do not change with the owner). Rubber tappers, however, knew that their stands were inside the reserve and that therefore their exclusive rights to use them were legal. These observations suggest that if need be, rubber tappers would engage in defending their rights against non-owners, but that they are unlikely to be managing the ERCM as their common-pool resource, since they do not know what their supposedly common resource is.

Representative organisations

By 1995, the three formal representative organisations of the ERCM had already been established[6]: the Associations of Inhabitants of the Extractive Reserve Chico Mendes of the Regions of Assis Brasil (AMOREAB), Brasileia (AMOREB), and Xapuri (AMOREX). These Associations were created during meetings with reserve inhabitants in 1994 (Ibama, 1995, p.21), and according to the Utilisation Plan they represent the reserve inhabitants, and are responsible for the management of the reserve. The rubber tappers, however, thought differently. They considered the rural unions rather than the Associations as their representative organisations. Many interviewees were unaware of the existence of the Associations and others referred to them as something that 'people from the city' and Ibama had established. Local leaders and key interviewees in general thought that it had been unnecessary to create the Associations, for the unions could have formally taken the role of representing the tappers in the management of the reserve – while some of these respondents may have had their own interests at stake, others would not necessarily have benefited from the unions having additional power.

One reason for the Associations' lack of representation among the forest dwellers could be that at the time of the interviews they had been in place for six months only. Three years later, however, in 1997, membership of the Associations was still low, limited to less than 50% of the reserve inhabitants (Irving and Millikan, 1997). This suggests that besides the time factor, there are also other reasons that explain the

Associations' inadequacy. One is that the Associations were created to address a necessity of which the tappers were unaware: the management of

Table 5.2 Organisations in the Extractive Reserve Chico Mendes

Rubber Tappers' Organisations	Level of Representation and Functions
Rural Workers' Union of Xapuri	Members in ERCM: 700 (out of 4000) Families in area: 848 Union branches in reserve: 13
Rural Workers' Union of Brasileia	Members in ERCM: 250 (out of 1200) Families in area: 370 Union branches in reserve: 8
Rural Workers' Union of Assis Brasil	Members in ERCM: 60 (out of 180) Families in area: 620 Union branches in reserve: 3
Associations of Inhabitants of the Reserve	Total number of members to date: unknown
Agro-extractivist Co-operative of Xapuri (CAEX)	Members in ERCM: 120 (out of 320) Warehouse branches in reserve: 6
National Council of Rubber Tappers (CNS)	Extractivist issues and negotiation of projects for extractive reserves
National Council of Rubber Tappers - Regional do Vale Acre-Purus (CNS-RVAP)	Responsible for matters of the ERCM
Federal Government Agencies	**Functions**
Ibama	Establishment and monitoring of reserves' regulations
CNPT	Technical and institutional support for reserves
Main Supporting NGOs	**Functions**
Centro dos Trabalhadores da Amazonia (CTA)	Education, health, community development, economic development

Sources: Instituto de Assuntos Culturais, 1993; GoB/BIRD/CUE, 1994; CNS, 1994.

the reserve as a whole. Forest dwellers had always used their resources in common without managing them in common. Yet, their resources had never been threatened because of fellow tappers and co-owners of the forest, only because of outsiders. The need to create an institution to manage the reserve was thus not felt, and with regard to certain functions of the associations that they considered important, such as solving sporadic disputes between neighbours, they thought the unions could do that.

Another reason is that the donors of the PP-G7 instead of the forest dwellers' representatives proposed the creation of association of inhabitants in the reserve – by and large the rubber tappers' leadership disagreed with this proposal. The argument for establishing the Associations was that the unions' constituency included many people living outside the reserve, and that their objectives were not limited to the sustainable management of the reserve. The unions' level of representation in the reserve is indeed debatable. As shown in Table 5.3, only a minority of the unions' members live in the reserve. The union's interests are thus unlikely to lie exclusively in the management of the reserve, and they may in some cases conflict with those of the reserves' inhabitants. Also, the level of representation of the unions inside the reserve is not comprehensive. Whereas in the reserve areas corresponding to the unions of Xapuri and Brasileia, there are 13 and 8 union branches, and most families have members in the corresponding union[7], in the area covered by the Rural Workers' Union of Assis Brasil, there are only 3 union branches, and out of 620 families, the union has only has 60 members. However imperfectly, the unions did represent the rubber tappers and their interests, and organisations focused exclusively on the management of the reserve could have been created within the unions' structures.

A third reason for the Associations' limited success is that forest dwellers participated little in the decision-making process that established them. This was partly due to the above two reasons, but also because of the way in which the Associations' representatives were chosen. The Associations were set up during meetings that took place outside the forest, in the cities of Xapuri, Assis Brasil and Brasileia. Many tappers did not participate in these meetings because travelling so far was too costly for them, and they were unaware of the importance and sometimes even existence of these meetings. In the case of AMOREAB, an additional problem had been that the organisers of the meeting, members of Ibama/CNPT and the unions, had insisted in choosing the representatives for the Associations in one afternoon. Because of such short notice, forest dwellers had accepted the leader proposed because they had not had the time to convince their own leader to accept the position[8].

Evidence collected from the study of other common property regimes suggests that the Associations may disrupt the organisational power of the Unions, without taking its place. In case of a conflict between neighbours, for example, a rubber tapper may initially seek arbitration from the union, and if unsatisfied with the union' decision on the matter, may then turn to the Association. If each organisation gives a different verdict on the conflict, it would be difficult to obtain a sustainable solution to it, as those involved would tend to recognise only the authority of the organisation that supported their claim.

The Associations present many shortcomings, but the 'genuine' rubber tappers' leadership is also flawed. Rubber tappers are divided in two groups: the direct users of the resource (forest dwellers) and the leadership or *lideranças*. The *lideranças* are not the same as the leaders of the various small communities in the reserve. The term leadership is used here (as it was used in the reserve) to refer to the leaders of the rubber tappers' formal organisations, such as the rural workers' unions, CNS, and CNS regional branch (CNS Regional do Vale Acre-Purus). *Lideranças* are in general tappers who participated in the process that led to the establishment of the reserve, and there is a marked difference between them and the general population. The former live most of the time in the cities and rarely work in the forest; they are literate and familiar with the reserve's bureaucracy, they know well the whole of the ERCM and its surroundings, and are self-confident in their dealings with outsiders. By contrast, the reserves' inhabitants seldom visit the cities, and are largely illiterate; they focus mainly on the management of their stands, and to a considerably lesser degree the rubber estate where they live – the reserve as a whole is largely ignored; in meetings with outsiders, especially if outside the forest, they hardly intervene in discussions[9].

The dichotomy between the forest dwellers and their leadership presents at least three important problems. First, there is the risk that the lideranças, who have more contact with outsiders, such as potential donors, and with CNS in Brasilia, channel the available financial resources to the rubber estates where they come from. This can lead to conflicts with rubber tappers from other estates that would feel unfairly treated; there is unconfirmed evidence that such situations have already arisen. Second, the leadership may favour family members to take up other leadership positions, and this can result in members of some estates regularly deciding on matters concerning other estates (the leader of the Rural Workers' Union of Xapuri in 1995, for example, was a relative of the previous Union leader). Third, as leaders stop living on the rubber estates and practising extractivism they cannot know the daily problems involved in the management of the forests (Ostrom, 1990). In creating the Associations of Inhabitants, two of the donors' objectives were to enhance the participation of forest dwellers in the management of the reserve, and break up the control some union members had of the reserve affairs. None of these two objectives, however, seems to have been achieved. The work load of an Association representative is such that the person in charge has little time left for practicing extractivism: one of them, for example, commented that he spent most of the time travelling to meetings with tappers from other estates, and had hardly any time to be in his seringal, let alone to tap rubber. With regard to breaking up the union veterans' control of the reserve affairs, in the Xapuri area this did not occur; the Association' representatives in Xapuri are formed by members of the union who live only part-time in the forest.

Rules, monitoring and conflict resolution mechanisms: the utilisation plan

The theory of common property argues that in robust regimes co-owners have devised rules specifically designed for the protection of the resource, which are appropriate to local conditions. The rules tend to be clear and easy to enforce, and there are institutional mechanisms for changing them. The environmental rules listed in the Utilisation Plan of the Chico Mendes Reserve, approved by Ibama in April 1995, were, in principle, drawn up by the forest dwellers. The Introduction of the Plan states that during the course of 1994, several community assemblies and meetings for all the reserve's inhabitants took place, during which the plan was drafted and discussed. The Plan also states that:

> [all the reserve] inhabitants are responsible for the execution of the Plan, as co-authors, co-responsible in the management of the reserve and the only beneficiaries of it. In a more direct way, the Associations of Rubber Tappers of the Extractive Reserve Chico Mendes, the rural workers' unions of Assis Brasil, Brasileia, Xapuri, and Sena Madureira and the CNS answer for the plan (Ibama, 1995, p.23).

The forest dwellers interviewed, however, thought differently: they did not consider the rules in the Utilisation Plan as their own. They believed the rules were 'natural rules' established by God, or alternatively, thought Ibama was the author of the Utilisation Plan. Two years later, according to a report from 1997, the majority of the reserve inhabitants had not yet internalised the Plan's rules, and continued to see them as made by outsiders (Irving and Millikan, 1997). The Plan of the ERCM is in fact based on the proposal presented by the representatives of the Alto Juruá Reserve (Feitosa, 1995), and although the rules were specifically designed to ensure the conservation of the common-pool resource and most of them are appropriate to local conditions in the Chico Mendes Reserve, they are critically flawed because they were not drawn by the ERCM inhabitants having in mind the protection of their common-pool resource.

Commercialisation of land According to the general legislation on extractive reserves, and to the specific rules of the Plan of the ERCM, tappers can interchange rubber stands with other extractivists (sell their usufruct rights), but cannot sell the land because the latter belongs to the state. The prices of the rubber stands are based on the number and quality of the rubber trees, and on the improvements *(benfeitorias)* stand-owners made to the original patch of forest, such as the agricultural plot and the number of rubber trails they cleared, the fruit trees they planted, and the house they built. The objective of this rule, that the reserve inhabitants can sell stands but not land, is to bar the forest dwellers from selling their stands to non-extractivists, and in this way to ensure the sustainable use of the common-pool resource by all the reserve inhabitants (see Chapter 4). The rule concerning land is thus specifically aimed at conserving the forest, and is clear and easy to enforce – as the tappers do not have private property rights they cannot sell them. It is also appropriate to local conditions because rubber tappers believe that land is not something that can be sold. According to most interviewees, land belongs to God or to Nature, and they can sell only what they have bought or what is the result of their work, as the improvements made to the rubber stands. Although stands have been traded in the rubber estates for many years, before the 1970s and the arrival of the cattle ranchers land was not traded as a commodity (see Chapter 2), and tappers never bought the land were they worked – the notion of land as marketable good was therefore an alien concept for the majority of the rubber tappers.

The problem, however, is that the prohibition to sell land was not a rule set up by the reserve inhabitants themselves. Apart from a small minority, rubber tappers were ignorant of the fact that it was because land belonged to the state that they were barred from commercialising it (in most cases they did not know it belonged to the state), and that this was partly a device to ensure the conservation of the forest. Tappers thus consider that land cannot be sold because of their economic, social and cultural background, rather than because they believe this is a good rule to ensure the protection of their common-pool resource.

Environmental rules Regarding the rules that deal directly with environmental issues, they are also based on the tappers' cultural background, but contrary to the case with the commercialisation of land, the reserve dwellers are fully aware of their importance for the conservation of the resource. Thus with regard to the ban on cutting rubber and Brazil-nuts trees (Utilisation Plan, Article n.8), for example, most interviewees said that this rule was appropriate because it was wrong to destroy what was not theirs – the trees belonged to God or to Nature. Yet, they also said that the trees should be conserved because they were important for their work. Another rule is that only up to 10% of the rubber stand can be cleared for agricultural or other purposes (Utilisation Plan, Article n. 15) (Ibama, 1995, p.25). As seen in Chapter 2, most of the Amazonian soils are very poor; if small patches of the forest are cleared for cultivation, the nutrients stay in the system and if the land is left fallow for a sufficiently long period, the forest regenerates. Although the majority of the forest dwellers interviewed were uncertain as to the percentage of clearing they were allowed to do in their stands, they knew why it was necessary to clear only a small patch of forest. They mentioned, for example, that in the agricultural settlements outside the reserve, where only secondary vegetation was left, the soil was less productive than inside the reserve: to obtain the same amount of produce a much larger area of land had to be cleared.

The rules stated in the Utilisation Plan are thus necessary for the conservation of the forests, they are clear, and they are known and approved by the tappers, hence adequate to local conditions. Those environmental rules whose utility is obvious to all tappers can in principle be changed if a change in the circumstances makes this necessary. With regard to the ban on the commercialisation of land, however, the tappers' unawareness of the importance of this rule for the conservation of the common-pool carries with it the risk of resource depletion. Traditional cultural values tend to be weak when faced with economic pressure and market values, and there are indications that this is also the case with the rubber tappers' values. According to Rueda (1995), upon the creation of the extractivist settlement projects, some tappers, influenced by the colonists, had wanted private property rights to their stands, rather than usufruct rights only. Also, if commoners do not know what a rule is for, they cannot rapidly change it to adapt to a change in the circumstances, which is a crucial requirement of robust regimes.

Monitoring mechanisms In addition to social pressure, robust regimes also have explicit mechanisms to monitor and enforce compliance with the rules. These mechanisms can be embedded in the rules or involve a separate set of arrangements. Monitoring tasks can be performed by the resource users or by contracted agents. In any case, however, the responsibility for ensuring compliance with the rules lies in the final instance with the co-owners of the resource. To guarantee enforcement, it is usually necessary to have sanctions against defectors, which should preferably be gradual. The Utilisation Plan of the ERCM appears to comply with all these requirements, since it says that the reserve inhabitants are responsible for monitoring the reserve, sets up different arrangements for checking observance of the established rules, and moreover outlines a set of gradual sanctions in case of non-compliance with the rules. The forest dwellers, however, are largely uninterested in monitoring their own use of the forest.

According to the Utilisation Plan, each individual tapper should monitor his and his neighbours' stands as well as form a Commission for the Protection of the Reserve, which should answer directly to the Associations[10] of Inhabitants of the Reserve (Ibama, 1995, p.27). The PP-G7 supports tappers' monitoring of their own resources by training them as environmental monitors. The Unions and Ibama are also responsible for monitoring. The rubber tappers, however, held that monitoring should be the responsibility of Ibama only. The generally held view was that it was wholly unpleasant to have to monitor other tappers' activities, and that Ibama officials could do this better because they live outside the rubber estates: 'the man from the state comes and goes' said one tapper when explaining why Ibama should monitor the reserve, 'while a member of the community has to go on living and interacting with the person who broke the law'.

Other tappers argued that monitoring was unnecessary since everyone complied with the rules. Partly because of social pressure as well as lack of incentives (if tappers clear their rubber trees they cannot continue tapping rubber, and there are hardly any available alternatives to earn a living), conformity with the established rules seemed indeed the norm. In the long term, however, incentives to break the rules will probably increase, and social pressure alone will prove to be insufficient to maintain the institutional arrangements. Moreover, although non-compliance is rare it does occur and social pressure has not always been able to deal with it.

Some rules have enforcement mechanisms embedded in them, like the one that prohibits selling of land, which cannot be broken because potential buyers need legal titles to the land that the tappers cannot provide. Many other rules, however, require separate monitoring mechanisms. If no one checks what rubber tappers in the reserve are doing, it will be easy for some tappers to sell their stands to loggers, hunt with dogs, and clear large patches of forest within their stands. Ibama alone cannot monitor the reserve. First, given the size of the reserve and the difficulty of accessing the rubber estates it would be very costly for it to do so. Second, even if Ibama was willing to invest whatever amount it was necessary in monitoring the reserve, it would be unlikely to succeed without the reserve inhabitants' full cooperation. Similar to what has happened in other cases in which state agencies tried to monitor a commonly used resource, potential free riders among the tappers could easily devise ways of eluding external monitors because they know the resource better.

The reserve inhabitants could monitor the reserve more effectively than any external agency. And, they could check compliance with certain rules with little extra work: as all the stands are intertwined, rubber tapers could in the context of their daily routine check whether someone hunts with dogs, or sells the stands to loggers; those tappers living near the reserve boundaries can similarly verify if strangers enter the reserve. Monitoring the size of the agricultural clearing, even of those tappers living quite isolated from their peers, would also be easier for the reserve inhabitants than for Ibama officials. What then explains the rubber tappers' lack of interest in monitoring? There are two fundamental reasons. One is that in spite of the efforts of government officials and leadership to convince the tappers of the risk that one tapper misusing his stand poses on all the others, many tappers continue to believe that the forest cannot be destroyed because of fellow tappers and that, therefore, monitoring is unnecessary. The other reason is that tappers who do believe it is important to monitor the reserve inhabitants' activities tend to think that it is preferable that others do it, because they are used to outsiders taking care of such matters, it is cheaper if others do it, and they are unaware that only they themselves can fully ensure compliance with all rules.

Tappers are used to manage only their own stand and to have external actors dealing with all other matters; first were the patrons, afterwards the middlemen, and more recently the Unions – in their perception, Ibama is now taking that role. The fact that they consider the rules are made by Ibama, also contributes to the thinking that Ibama is responsible for these rules. Then there is the cost factor. Monitoring the reserve and enforcing compliance with the rules would be costly; tappers would have to take time off work to monitor tappers who live far away, and they could have problems with their neighbours. From the tappers perspective, it is thus cheaper to have Ibama official monitoring the rules than to do it themselves, even if in absolute terms is cheaper for them than for Ibama. The cost of enforcing the rules would diminish, however, if everyone agreed that they needed to be enforced – the enforcer would be appreciated by the group, and it would be the infractor that suffered social condemnation. Tappers could also devise a scheme by which they would be only responsible for monitoring instead of doing the monitoring themselves. Finally, the benefit of monitoring the reserve, which is the conservation of the forest, would probably offset the costs of monitoring and enforcement.

Conflict resolution mechanisms There are no specific conflict resolution mechanisms in the Utilisation Plan, and in case of a dispute between neighbours is unclear which of the several organisations with responsibilities in the reserve – Unions, Associations, Ibama – would have the authority to give a final verdict. The forest dwellers have non-institutionalised incipient conflict resolution mechanisms: when a conflict arises they usually discuss it between themselves, in small groups of 5 to 6 families and seek the opinion of one of the local leaders. In case of a dispute they cannot solve between themselves, however, they lack any institutional structure within the forest to address the issue and seek instead the help of an external organisation because they deem that outsiders would also know better how to resolve the question, and would have more authority. The problem, however, is that not all tappers seek the same organisation: depending on which organisation has the stronger presence in their area, they refer to either the local union, Ibama, one of the Associations, or the police. A decision by any of these organisations could therefore be questioned by referring to another organisation, which would render the resolution of any conflict extremely difficult.

To summarise, as a common property institution the reserve is not robust. The boundaries of the resource and of the group of co-owners are formally well defined, but largely unknown to the co-owners themselves; there are rules specifically aimed at the conservation of the forest, but they were not designed by the forest dwellers; monitoring and enforcement mechanisms are weak, and conflict resolution mechanisms virtually non-existent. Most importantly, forest dwellers participate very little in the management of the reserve, either directly or via their representative organisations, and show hardly any interest in doing so in the future. This situation, however, cannot be blamed on the state or on the leadership wanting to take control over the tappers' forests. The objective of state agents and former rubber tappers is clearly the management of the reserve by its inhabitants. Excepting some articles from the Brazilian environmental legislation, most rules in the Utilisation Plan are the same ones tappers had before the establishment of the reserve[11]. All formal documents of the reserve such as the Utilisation Plan, legislation on extractive reserves, reports for the Pilot Project, emphasise the forest dwellers are co-responsible for the management of the reserve, and there are many initiatives to enhance the tappers' capacity to manage their resources, such as training of environmental monitors. At meetings, state officials and lideranças try to encourage the forest dwellers to take over the management of the reserve, by telling them they are the owners, explaining why is it necessary that they become responsible for monitoring the forest, and also by providing training in various skills that are necessary to manage the reserve.

The research carried out in the area suggests that the core reason for the tappers lack of participation in the management of the reserve may be that forest dwellers have a different understanding of their space and institutions from that of the *lideranças* and state agencies. (Other reasons will be discussed at the end of the chapter.)

For outsiders the relevant management unit is the reserve. Accordingly, they have established rules that apply to all the reserve, and the emphasis on meetings between tappers and outsiders is always on the reserve as one single unit: 'This is your reserve and you have to manage it together'. For the forest dwellers, however, the relevant management unit is not the reserve; for them it is the rubber stand, and to a less extent the rubber estate where they live or, if the estate is too large, a part of the estate. Rubber tappers consider the inhabitants of estates located elsewhere in the reserve almost as foreigners: a remark they often made, for example, was 'on estate so and so they are all Indians', meaning people there were different from them[12], and a tapper from Icuriã estate who had been living for several years in neighbouring Guanabara estate, mentioned once that she only socialised with her previous friends from Icuriã because she found people in Guanabara too strange. From the forest dwellers' perspective, the common-pool resource is thus not the reserve, nor the inhabitants of the reserve 'their community'; for them, the shared resource is the rubber estate where they live, and 'their community' is formed by their neighbours, with whom they have relatively frequent contact. They know their rubber estates happen to be located inside the ERCM. For them, this means that the ranchers cannot occupy their rubber stands, and that they have to follow 'Ibama laws' in the same way that they had to follow the patrons' laws before; it does not mean, however, that they are entitled to manage the reserve and are responsible for its conservation. In other words, from the forest dwellers' perspective the reserve is outside their jurisdiction.

To ensure the conservation of natural resources in the ERCM it is thus necessary to overcome the following problem. On the one hand, given the characteristics of the forest, a robust regime requires the conservation of the entire reserve, and hence the development of a common property regime involving all the inhabitants of the reserve. But on the other hand, given that the reserve inhabitants regard as the relevant common-pool resource only the rubber estate where they live, and show little interest in the inhabitants of other estates, they are unlikely to set up an institution that comprise the entire reserve. One way of solving this problem, however, is to develop a system of nested enterprises (Ostrom, 1990) based on the articulation of common property regimes on the different rubber estates that compose the reserve. The research conducted on five rubber estates suggests that the potential for robust regimes to develop within the *seringais* is higher than for the reserve as a whole. Once tappers develop small regimes, the need to articulate the use of their estate with the use of neighbouring estates is likely to become more apparent. The state and lideranças can then encourage arenas for discussion among representatives of the various estates instead of promoting the management of the entire reserve as a single unit, which is, as mentioned above, beyond the perceived jurisdiction of the forest dwellers[13].

PART II – THE RUBBER ESTATES

Common property regimes on the rubber estates – internal factors

This section argues that the likelihood of a common property regime developing within a rubber estate is higher than it developing for the whole reserve because in the rubber estates there are incipient common property regimes that have a strong potential of becoming robust – many of the factors highlighted in the theory as conducive to the development of robust regimes could be discerned on the *seringais*. By contrast, at the level of the reserve, the only institution in place is the formal co-management regime, which the forest dwellers did not develop, in which they show little interest, and where there are fewer factors conducive to the development of robust regimes.

The rubber estates' communities have developed incipient common property arrangements to address three issues: use of common areas (e.g. forest paths, lakes and streams), sharing of common facilities (e.g. warehouses, schools and health posts), and commercialisation of their produce.

Forests paths can be jointly used by all those who need them, but they need to be cleared regularly, in general two or three times a year, otherwise the fast growing vegetation covers them, and they become extremely difficult to walk on. Communities tend to meet regularly to clear the paths, but there are no rules specifying when the paths should be cleared, and who should participate in this activity. Sanctions against uncooperative community members are virtually non-existent. Communities tend to dislike non-participant members, but they do not engage in ostracism specifically because they fail to help clearing the paths. Everybody, including non-cooperative tappers and outsiders can freely use the paths.

Institutional arrangements regarding common facilities, such as schools, health posts, and warehouses, ranged from non-robust common property regimes to restricted access. Community Primavera, on Seringal Icuriã, exemplifies the first case. Its members had formed a co-operative for commercialising their produce and buying consumer goods in the city.

They own, in common, animals for transporting goods between the rubber estate and the city, a pasture to feed the animals, and a warehouse. Only the formal members, who pay a quota, can commercialise their goods through the co-operative, and every two years they elect a leader, which is also in charge of transporting the produce to the city and bringing back consumer goods; every two weeks the members gather to clear the pasture together. This was one of the most well-structured institutions found inside the reserve, but it was relatively weak because apart from social pressure, the only sanction they have is withdrawing the use of the co-operative in case the person fails to pay too many of their quotas. On rubber estate São Pedro, however, incipient institutions were considerably weaker than Primavera, since both the boundaries of the group of co-owners and their responsibilities were ill defined; this may have been because the regime was very recent. There are several common facilities on the estate, which outsiders provided, and that in principle belong to all the inhabitants of the estate. There are, however, two communities within the estate, Itapiçuma and da Margem, and they disagree on who the co-owners of the facilities are. According to Community Itapiçuma, the pond, the vegetable plot, the radio, the warehouse and the animals belong to both communities; according to Commmunity da Margem, these facilities belong to Itapiçuma only. The former complain that Community da Margem refuse to participate in the common activities, and Community the Margem argues that they would participate if they also had their own facilities.

Although common property institutions on the rubber estates are at an incipient stage, they have considerable potential to develop further and become robust regimes, capable of ensuring the sustainable use of the forest in the long-term. This is because 1) the estates inhabitants tend to form 'communities'; 2) the boundaries of the estates are well known to all their inhabitants; 3) estate communities have monitoring mechanisms; and 4) communities are aware of the need to manage their estates.

Whereas the 'reserve inhabitants' form a large and heterogeneous group, and there is little contact among families living in different parts of the reserve, families on the rubber estates tend to form small and relatively homogenous groups, with a more or less stable set of members who expect to continue interacting over time, have relationships which are direct and multiplex, and are mutually vulnerable actors (Singleton and Taylor, 1992). Chapter 1 argued that these factors alone would not result in the development of robust regimes, but that all other factors being equal, resource users meeting Singleton and Taylor's (1992) requirements were more likely to develop a regime than a large and heterogeneous group. All other factors are mostly equal for both the individual estates and the reserve: the attitude of the state, the users' dependency on the resource, their knowledge of the forest, and their level of autonomy are the same for the 'estate communities' and for the 'reserve inhabitants' (for the rubber estates are part of the extractive reserve). Hence, given the features of the inhabitants of the estate, robust regimes can more easily develop within each estate than at the level of the reserve.

Most forest dwellers considered themselves as belonging to a community formed by 4 to 25 families living in the same area (see Table 5.4). The size of the area varies, but generally, families are, at most, 5 hours walk from a meeting point. Communities are formed either by all the families of the estate or by a sub-group of families; in the latter case, the estate represents the wider community.

The families forming each of the estate-communities are quite homogeneous. They all practice the same activities, be it rubber tapping or collection of Brazil nuts, and have a similar history; for example, all had the same patron in the past, or had to face a similar degree of risk in relation to ranchers occupying their rubber stands. The communities have a more or less stable set of members: most families have been living in the same estate for generations[14]. Relationships among families are direct, and in spite of the distance between houses, forest dwellers of the smaller communities are frequently in contact with each other. Some relationships are multiplex, as in the case of communities Primavera and Itapiçuma whose members meet for work, leisure, and religious activities; in other cases relationships are based on one type of activity only, such as the organisations of balls, or the gospel reunions. The 'mutual vulnerability' of families varies: the respect and friendship of the other members of the group is important on those estates where families have regular contact; on those, like Guanabara, where they seldom see each other, it is largely irrelevant.

The potential of rubber estates to be managed as common property is higher than that of the reserve because seringais are already held as common property, and all forest dwellers know the boundaries of the *seringal* and its inhabitants (the co-owners). Hence, tappers would know what to manage and for whom. Regarding the jointness conditions of the estate, they are similar to those of the reserve and thus conducive to the development of common property. The resource can be jointly used providing the activities practiced do not require clearing the forest cover[15].

Table 5.3 Communities on the rubber estates

Rubber estates	Communities
São Pedro	São Pedro – 25 families Itapiçuma – 10/11 families Da Margem – 14/15 families
Icuriã	Primavera – 24 families Rice Mill – 4 families Gospel communities – 5 families each
Dois Irmãos	Dois Irmãos – 17 families Gospel communities – 5 families each
Floresta	Rio Branco – 5/15 families
Guanabara	No indication of community

Whereas forest dwellers had hardly any interest in monitoring the use of the reserve, they were quite concerned with monitoring the use of common facilities on the rubber estates, and had developed mechanisms to deal with this issue. The location of facilities is not based on a spatial differentiation between 'common areas' and 'private areas', but on the distribution of resources on the estate, and on the configuration of the estate in terms of stands and inhabitants. This arrangement facilitates monitoring. Schools, health posts and warehouses are located in private rubber stands and thus the owner of the stand can easily monitor access to them without additional work. In addition, the stands where the facilities are located tend to be meeting places for the community or 'cross-roads' where everybody tends to go for other motives. The members of the group can thus also regularly check the use of the facility in the context of their current activities. The generally held view of the interviewees was that common facilities had to be located in the stand of the leader of the community or a long-standing member of the group, someone everyone trusted.

And, finally, another factor that indicates that tappers can develop common property regimes is that when they perceive the need for institutionalised cooperation they engage in it. Families have for example developed institutionalised cooperation for clearing the forest paths, which is something they clearly perceive a need for. As soon as vegetation begins growing they become unusable and seriously restrict the tappers' mobility. The more families need the paths, the more effort they put into maintaining them in good condition and actively ensuring that no one obstructs the paths. Hence, on estates were the commercialisation of the produce is done by a cooperative, tappers get regularly together to clear the paths, whereas in those areas where the middlemen do it tappers are largely unconcerned with the state of the paths – the one that uses them more is the local middleman.

To participate in communal activities, tappers must perceive the need for them, and also believe that the advantages they will gain from working in something for the community will offset the cost that represents losing work hours in their own rubber stands[16]. More tappers participate in a communal work that brings perceived advantages to many of them, than in one that will benefit only a few: more people will participate in the construction of a bridge that facilitates everyone's access to the nearby city, for example, than in the construction of a school useful only for those living at most two hours away from the site. The advantage gained from participating in a common endeavour can also be receiving help later. Hence, in case of sickness – taking someone in the hammock to the nearest city – everybody helps, since everyone may need help in this respect. Tappers also regularly help each other clear their agricultural plots, since the work involved is clearly too intensive for a single family. When the yearly clearing of the agricultural field needs to be done, the family calls the neighbours, and they all work together; the owner of the plot that was cleared will owe 'work days' to those who helped him, and will repay by working in their private fields at a later date[17]. Although there are no detailed rules, social pressure and the need for reciprocity makes it necessary for tappers to help their fellow tappers on their plots.

An argument put forward in Chapter 1, was that for common property regimes to develop it was first necessary that commoners thought necessary to cooperate and institutionalise cooperation. This was also the view of the forest dwellers: they thought that those who refused to participate were unaware of the improvements that common facilities could bring to their lives. Hence, instead of refusing 'free-riders' the use of their facilities, they encouraged uncooperative tappers to use them, for they thought this would help them realize the advantages of working together, and cooperate in future communal works. This strategy had already showed good results. The large majority of the forest dwellers are unaware they need to develop mechanisms to ensure the conservation of the forest, but once they perceive this need, it can thus be safely assumed that they will create the necessary institutions.

This section has explored the potential for the development of robust common property regimes on the rubber estates focusing on 'internal factors'. It has shown that tappers have engaged in various incipient institutional arrangements, but that they have done so only when the need for co-operation was clearly perceived. This section has also argued that most internal factors conducive to the formation of common property were discernable on the rubber estates. Families form groups which meet most features that enhance the co-owners' capacity to set up resource management systems; and the estates have jointness and exclusion conditions that permit holding them as common property and that facilitate monitoring the use of the forest. In other parts of this volume, it has been shown that tappers are highly dependent on the common-pool resource, and that they know the ecological limits of their forests, but that they are largely unaware that they need to actively regulate the use of the forest by their fellow tappers to ensure the conservation of their own private rubber stands. All these factors suggest that tappers *can* develop common property regimes on the rubber estates, and that one of the main reasons why in fact they have not done is that they think is unnecessary.

Common property regimes on the rubber estates – the external context

The possibility of common property regimes developing within the rubber estates depends also on 'external factors'. The reason why tappers used their forests for such a long time jointly and sustainably, but cannot continue doing so without strengthening their common property regime is that the external context has changed. In the 1950s and 1960s, the rubber estates were largely inaccessible to outsiders, who had little interest in these areas unless they wanted to tap or trade rubber. Hence, it was unnecessary for tappers to protect the *seringais* against outsiders or to develop mechanisms to ensure that fellow tappers sold their stands only to extractivists. Presently, the situation is different: loggers, for example, may attempt to enter the non-inhabited areas of the reserve, or buy isolated stands from tappers wanting to leave the forest. What is more, tappers themselves may decide to switch to destructive activities such as logging for which previously there was no market.

How the rubber tappers can deal with such a change in their circumstances is correspondingly influenced by external factors.

Particularly important is the level of autonomy external actors leave them and the support they offer them. Equally important are the level of contact the tappers have with the outside world, and how external factors enhance or constrain tappers' access to information. (It should be noted that the focus will be on the impact of external factors on the tappers' capacity to harmonise their *own* use of the resource, and not on their capacity to protect the resource from outsiders, an issue which was already explored in Chapter 3.)

Throughout their history, tappers have had little autonomy to manage their resources, and to take any type of initiatives. This, however, and contrary to most other cases in the common property literature, has been unrelated to state interference; their autonomy has always been constrained instead by powerful private agents. During the aviamento period the rubber barons and the patrons forbid tappers to develop their own arrangements regarding use of the forest. With the different crises of the rubber trade, the tappers acquired some autonomy to manage their resources; taking time off extractivism for cultivating a garden plot, for example, was a decision that tappers could take only after the barons left the estates, and once the patrons' hold on them was weakened, for during the rubber boom they generally banned them from practicing agriculture. The middlemen have also had an important role in curtailing the tappers' autonomy[18]. Given the distances between houses and the isolated character of the tappers' work, many forest dwellers have more regular contact with the local middleman than with each other, and middlemen have often advised the extractivists against any form of association saying that it would make them lose their stands[19]. Besides, by keeping the forest dwellers always in debt, middlemen have kept rubber tappers in a type of psychological dependency that holds them back from rebelling and setting up their own trading arrangements.

The tappers' historical lack of autonomy helps to understand why now, when given the opportunity to manage their own resources – at least to a certain extent – forest dwellers keep from taking full advantage of this opportunity. By and large, and apart from decisions directly concerning the rubber stands, tappers leave outsiders to make all suggestions regarding the management of the forest. To speak up at meetings where outsiders also participate forest dwellers need a great deal of encouragement by the outsiders and their own leaders. Hence, decisions are often taken by outsiders (which include the *lideranças*) rather than by the forest dwellers. Another reason why forest dwellers shy away from taking initiatives, especially in the presence of strangers, is that they believe outsiders are more knowledgeable than they are. Low literacy rates, isolation from the rest of the world, and little access to information (the only two means by which tappers acquire information are verbal communications by people who have been outside the forest, and radio, since without electricity they cannot have television) contribute to inhibit their self-confidence, especially regarding issues of which they have little previous experience.

Within the institutional structure of the ERCM, the tappers' autonomy is relative. On the one hand, the reserve co-managers such as Ibama officials and some lideranças, try to encourage the reserve inhabitants to participate in all decisions regarding the reserve, and in particular in those which concern their own rubber estates. Hence, for example, forest dwellers are the ones that choose from a range of options which facilities to have on their estates. The formal co-management system of the reserve also leaves the tappers sufficient autonomy to develop rules and monitoring mechanisms at the level of the rubber estates. The only stipulation made about the use of common areas in the ERCM Utilisation Plan is that the community is responsible for them and for stipulating how they should be used (Ibama, 1995, p.26).

On the other hand, however, the co-managers of the reserve also hinder the tappers' autonomy. The rubber tappers have little experience of managing their resources in common – they are moreover uncertain that is necessary to do it and lack the self-confidence to initiate it. By suggesting initiatives, and afterwards encouraging tappers to implement them before they have had the time to think about them, and to develop these new ideas in ways that fit the specific characteristics of the estate where they live, external actors impair the tappers' level of initiative. The formal co-management system of the ERCM also restricts the autonomy of the inhabitants of the rubber estates in important ways. The following examples illustrate this point. According to the Utilisation Plan of the reserve, forest dwellers can change all the rules that govern the reserve apart from those established by the Brazilian legislation. Yet, a proposal to change a rule must be presented by at least 10% of the families inhabiting the reserve, which corresponds to approximately 200 families, and the proposed change needs to be approved by a minimum of 400 families. As the groups in the rubber estates tend to be formed by at most 25 families, it is next to impossible for forest dwellers to change a rule. Another case in point is the monitoring system of the reserve. By encouraging tappers to become monitors of the entire reserve this system can hinder the development of more effective mechanisms at the level of the estate, where some incipient arrangements already exist, since forest dwellers may get used to rely on tappers from other estates, who from the tappers' perspective are like outsiders, to ensure the conservation of their resources. This would defeat the purpose of having rubber tappers monitoring the reserve. A group of tappers who thought the existing monitoring system ineffective would need to convince all the inhabitants of their estate and of several neighbouring estates to be able to propose a change.

External help has always been a determinant factor in the development of common property on the rubber estates. One of the tappers' main problems is their dependency on the middlemen – most forest dwellers referred to this issue without even being asked about it – yet cooperatives for the commercialisation of rubber only developed on those estates where an external agent encouraged the tappers to set one up. Less knowledgeable and powerful than the middlemen, the tappers were largely unable to break free from the middlemen's control without outside support. This support has included financial means (e.g. to buy animals to take the produce out of the forest), encouragement and provision of information (e.g. about how to form a cooperative). External help also explains why some groups developed incipient common property arrangements and others did not, although as communities they are all essentially similar – the same number of families and the same degree of homogeneity. The only difference observed between such communities was the level of external presence on the rubber estate.

Outsiders have brought in most common facilities on the estates, namely schools, health posts, and warehouses; outsiders also encouraged forest dwellers to undertake common works such as clearing the forest paths or building a bridge, and to create cooperatives. Particularly important was the presence of members of the Catholic Church. Most of the communities on the estates were initially Christian Grassroots Communities (see Chapter 3), and all the tappers' leaders, both those living in the forest, and those belonging to the leadership, learned how to read, write, and the basics of community organisation thanks to Catholic priests and nuns. The importance of the church in extractivist areas can be exemplified by comparing two neighbouring rubber estates of Icuriã and Guanabara, which although similar in many respects, are institutionally very different. In Icuriã, there is a local cooperative and households maintain regular contact and seek each other when faced with a problem. By contrast, in Guanabara, the forest dwellers are totally dependent on the middlemen, they hardly visit each other, and are considerably poorer than their neighbours on Icuriã. A fundamental difference between the two rubber estates was that a priest had lived for many years on Icuriã and had only sporadically visited Guanabara; on Icuriã he had been responsible for setting up the cooperative and had encouraged the tappers to clear the forest paths.

Also very important were the Rural Workers' Unions and non-governmental organisations, mainly CTA (*Centro dos Trabalhadores da Amazônia*, Amazon Workers' Centre), founded in 1983 with the purpose of supporting rubber tappers. CTA developed primers specifically designed for children living in the forest, and in co-operation with the Union of Xapuri set up schools on the rubber estates. CTA also helped establishing basic health posts on the estates, providing training as health agents to the tappers, and undertaking extension work in the forest. With support from international organisations, CTA and the Rural Unions established in 1989 a Brazil-nut factory at Xapuri, which developed in a co-operative, the Agro-Extractivist Co-operative of Xapuri (CAEX). In 1993, this co-operative had 6 warehouse branches on the rubber estates and 120 associated members inside the reserve (ELI, 1994)[20].

Currently, Ibama and the RESEX sub-project are the main supporters of the ERCM, providing education and health services, extension services and encouragement in the establishment of cooperatives as well as in the development of cooperation in general. Tappers are encouraged to have meetings, they are provided with material means and information, and external actors travel regularly to the reserve to encourage tappers to participate in common activities, e.g. to build a bridge or a school. Although Ibama's attempts at establishing institutional arrangements in the reserve are problematical, the help they provide otherwise has been crucial for the improvement of the tappers' lives in the last few years, which could not have obtained by themselves, for example, teachers specialised in various fields, and construction materials.

Whereas for other commoners isolation was what permitted them to develop and sustain robust common property regimes, the opposite occurred on the rubber estates. There are, for example, more common property arrangements on rubber estates that are easily accessible than on remote ones. By and large, the estates can only be reached on foot; to reach some of them it's possible to go most of the way on boat, but once inside the estate it is necessary to walk, since there are very few animals that can transport people (the few animals available are used for transporting goods and for medical emergencies). Some estates, however, are considerably more isolated than others: Dois Irmãos, for example is three hours walk from a road which connects Xapuri with Rio Branco, the capital of the state of Acre, whereas Guanabara is one day walk from the nearest city, Assis Brazil, which moreover is inaccessible by car during the six months of the rainy season.

Accessibility is a positive factor for the development of common property regimes on the *seringais* because it facilitates contact with the outside world and enhances socialisation between rubber tappers. The closer a rubber estate is to a road or to a city, the more likely it is that outsiders would have been there. In addition, the most accessible rubber estates were the most affected by the invasions of the cattle ranchers. Hence, in these areas more people participated in standoffs and benefited from all the experience of co-operation that such movements brought with them. Rubber tappers living on estates which are only a few hours away from a market centre can more easily take the time to travel there to commercialise their produce; hence, they are less dependent on the middlemen than tappers who need to travel for one day to reach a market place. Moreover, travelling to a city increases their access to information, diminishing their reliance on the news brought in by the middleman, and augment their possibility of meeting other tappers, either on the road or in the city, and of discussing issues of common interest. By contrast, a tapper living two or three days away from a market centre cannot afford to lose so many workdays to take the produce to a market centre; therefore he will have to sell his rubber to a middlemen. This has several implications: he will have to work harder, because the prices paid by middlemen tend to be very low, and so will have less time to participate in common works than the tappers who can sell his produce in the city. Moreover, he will have considerably less access to information, since time to attend meetings and visit neighbours is scarcer and the possibility of meeting people from outside the state rare. In most situations, he would have to rely on the biased version on the middlemen for outside news.

Until recently, articles about common property institutions seldom mentioned leaders as one of the crucial factors for the development of common property (Ostrom, 2001). They are, however, considered very important in other studies, and on the rubber estates, they were fundamental for promoting cooperation and institutional arrangements between the rubber tappers. On rubber estates where incipient common property institutions exist, tappers usually have a leader, whereas on estates where no cooperation between the forest dwellers has developed, there is generally no established leader. The leaders of the rubber estates had one important characteristic: they all had more contact with the outside world than the other inhabitants of the estate. Some had lived in the cities for some time and others still lived there but now only part-time[21]. Most of the leaders travelled regularly to the nearby city. Often the leader was someone who had worked closely together with outsiders on the estates, either members of the church or union leaders. Contact with the outside world provided the leaders with more knowledge, with better access to information, and with the capacity of establishing a link between the tappers and the external world.

One of the variables that need to be assessed when exploring commoners' capacity to develop robust regimes is what is their access to information (see Chapter 1). This will depend on both internal and external factors. For instance, the simpler the resource and the more educated the resource co-owners, the easier it would be to access information that will permit the development of a management institution. In the case of the tappers it was also observed that access to information gives self-confidence to set communal arrangements; one of the reasons why tappers keep silent at many meetings is that they believe that outsiders know better and have access to more information. The leaders of the estates are also those that have more access to information. The external context can restrict or enhance commoners' access to information. As noted earlier, middlemen have often restricted tappers' access to information by providing them with biased information; by contrast, church and union members, as well as NGOs and Ibama have provided forest dwellers with valuable information on better production techniques and on how to set up co-operatives.

To sum up, the forest dwellers have more access to information and more external help than at any time before, two factors that by themselves should hasten the development and strengthening of robust regimes on the rubber estates. Their lack of autonomy, however, past and present, seems to be hindering this process. Implementing institutional arrangements in the reserve without the tappers having felt the need for them, and which are inadequate to the local conditions of the rubber estates, is unlikely to promote the development of robust regimes in the reserve. First, given the tappers' lack of experience regarding the common management of their resources, and that they have been used to outsiders managing their resources, the approach of external actors can just result in the tappers transferring their relationship with the patrons to the new outsiders: to a certain extent this is happening already. Second, the approach of external actors has led tappers to believe that Ibama and the lideranças manage the reserve, in which case it is unnecessary for them to incur the costs of managing the common-pool resource, they can just use the resources for their own individual benefit. Third, although tappers do not voice their complaints, they easily ignore outsiders' suggestions with which they disagree - unless there are effective enforcement mechanisms in place such as those of the *aviamento* system.

PART III – OTHER EXTRACTIVE RESERVES

In addition to the Chico Mendes Reserve, there are another 32 extractive reserves in Amazonia, 11 of which are federal reserves, and the rest state-administered reserves. It is important to have an overview of other extractive reserves because it provides a source of comparison with the case study. There are, however, also two other reasons pertaining to the themes of this book: autonomy and external context.

The theory on common property institutions stresses that it is important for commoners to have sufficient autonomy to manage their resources and the analysis of the Chico Mendes confirms this. In comparing the ERCM with other reserves, the need for commoners' autonomy also becomes apparent. Although all reserves are inhabited by individuals who hold their resources in common and depend on the forest for their survival, each of them presents sufficient specificities to make it virtually impossible for an outsider to design an institution appropriate to the local conditions. In the same way that private property takes many different forms – the rules that govern, for example, the use of private houses in two different countries are rarely similar – common property institutions also vary substantially, because specific jointness and exclusion conditions of a resource are different. As a group, resource users also differ: commoners have different histories of collective action and different levels of homogeneity.

Comparing the ERCM with other reserves also serves to note an important feature of the interaction between local resource management and the external context: even if all commoners share the same external context, the external factors that will influence each group of commoners can differ substantially. The forest dwellers of all the reserves reviewed here are commoners who, like those of the ERCM, have been affected by the crises of the rubber trade and the government policies of the 1970s and 1980s (see Chapter 2); and since the establishment of the reserves, they have all been influenced by the action of Ibama. The interrelationship between local institutions and the external context, however, has been different for each reserve. Although the external context is the same for all extractivists living in a region, each group is affected by a different part of it. Thus, for example, whereas the inhabitants of the ERCM were affected by the arrival of cattle ranchers, those of the Rio Ouro Preto Reserve in Rondônia were mainly influenced by the Polonoroeste project – nonetheless, both factors, stem from the government policies for Amazonia in the 1970s (see Chapter 2).

The next few pages provide a brief overview of the situation in two other Amazonian extractive reserves: Alto Juruá, in the western part of the state of Acre, and Rio Ouro Preto in neighbouring Rondônia. Like the Chico Mendes Reserve, all of them were created in 1990, are federal reserves managed by Ibama, and have been supported by the PP-G7 since 1995.

Alto Juruá

In contrast with the rubber tappers of the Chico Mendes Reserve, the inhabitants of the Extractive Reserve Alto Juruá (ERAJ) never lived as autonomous tappers: until shortly before the establishment of the reserve, the *aviamento* system predominated in the region. It was not as violent as during the rubber boom, but patrons maintained the characteristic debt-bondage ties with the tappers. The government approach to the Amazon region during the 1970s and 1980s also affected the tappers living in Alto Juruá, but the policies that affected their lives were different from those that affected the inhabitants of the Chico Mendes Reserve. The main external impact in the Alto Juruá Region was the government rubber policy: when in 1985, the government stopped subsidising rubber, patrons decided to switch to logging and stop the credit system of *aviamento*; in 1986, supported by the local police, they went to the rubber estates to collect their debts. This led to a revolt on the rubber estate Restauração, in the Rio Tejo area. The first major incidence of collective action on the rubber estates of Alto Juruá was thus not against outsiders trying to take over the common pool resource, but against their own patrons. The similarity between the struggles of the tappers in Alto Juruá and Chico Mendes is, however, that in both cases they had to fight against private agents, which were more powerful than they were.

At the time the tappers in Alto Juruá revolted, the wider socio-political context was more facilitative than when the inhabitants of Southern Acre began their fight against the ranchers. International NGOs and the public in industrialised countries, and in the more developed regions of Brazil were aware of the existence of the rubber tappers, since one year earlier, in 1985, the National Congress of Rubber Tappers had taken place in Brasilia. There were also incipient institutional arrangements to support the tappers. The National Council of Rubber Tappers, for example, had been already set up and its mandate was to help extractivists all over Amazonia; their representatives visited the Alto Juruá region soon after the tappers' revolt, and in 1988, thus two years after the revolt, the idea of setting up an extractive reserve in the area was vented. At the time the tappers of Southern Acre revolted, they lacked any specific solution to their problems because none was available in the existing legal framework of the country. What is more, their first struggle was to make their existence and plight known in the rest of Brazil and the world.

The history of the forest dwellers in Alto Juruá – a longer history of dependency and less experience of collective action the tappers from western Acre – has influenced their group characteristics. Overall, the development of communal institutions in the ERAJ is particularly difficult. The ERAJ is divided in two areas: one near River Tejo, which is where the revolt against the patrons took place, and which is the most developed area in terms of community organisation, and one near River Juruá, where community organisation is virtually non-existent. There is one single organisation to represent both groups, the Association of Rubber Tappers and Small Farmers of the ERAJ (initially named Association of Rubber Tappers and Small Farmers of the River Tejo Basin). This association was set up in 1989 by the tappers' representatives, unions and CNS, rather than by external actors as the ERCM's Associations, so it can be expected that it is more representative of the forest inhabitants than the latter. A question that arises, however, is whether having one single organisation to represent two very different groups, and which initially seemed to represent only one of them, those of the Rio Tejo Basin, is conducive to the development of cooperation. It may be that the Association attracts tappers with little collective action experience, or that it hampers the development of communal institutions because half of the reserve's inhabitants feel left out.

Evidence on the success of the Association is inconclusive. On the one hand, and according to Ibama, participation in the Association appears to have doubled in the last five years: in 1995, it amounted to nearly 40% of the population (Feitosa, 1995), but in 1999 to over 80% (Ibama, 1999). On the other hand, attempts by the Association at setting up a cooperative for the reserve have failed. This was partly because middlemen continued operating in the area and tappers continued buying for them. The middlemen maintained the best part of the market: while the cooperative commercialised products which were low value and heavy, such as ammunition, soap, salt, and work tools, the middlemen traded on 'luxury' goods, expensive and low in weight. Another problem was that the forest dwellers expected the newly founded cooperative to give them long-term credit as the patrons had always done. But given the precarious situation in which they were at the time because of the low price of rubber (an external factor), and the absence of heavy coercive methods they were used to (an internal factor) they failed to pay their debts on time (de Almeida, 1994).

Another difference between the two reserves is that whereas in the ERCM tappers tended to be grouped in single families only – so each family lived from a considerable distance from other families – in the ERAJ tappers tend to form groups of 6 to ten families, all living in the same stand, but in different houses. From this, it could be concluded that local communities in the Alto Juruá reserve are likely to be more cohesive than those in the ERCM, but then again, given all the other characteristics of the tappers in the area, such as lack of experience as autonomous tappers, this may not be the case.

The exclusion conditions of the ERAJ area are different from that of the Chico Mendes Reserve. The ERAJ is a large and complex resource, but its area (approximately 5, 000 sq. km) is nearly half that of the Chico Mendes Reserve and its population (less than 6000 inhabitants) is likewise considerably smaller. The boundaries of the reserve were demarcated in 1995, and it can be assumed that protection against outsiders is easier in Alto Juruá because the reserve border areas where the standing forest is also valued – a natural park and four indigenous areas, inhabited by the Kampa, Jaminawa-Arara, and Kaximinaua (de Almeida, 1994; Feitosa,

1994). Besides, given the inaccessibility of the region - not accessible for 10 months of the year – it is unlikely that newcomers would attempt to take over the area. On the other hand, the existence of many waterways in the region has facilitated the entrance of loggers in the neighbouring areas, and the same can happen in the reserve.

It is beyond the scope of this study to assess the institutional differences between the two reserves, but the dissimilarities of the commoners suggest that to be effective, their boundary and harmonisation rules, as well as their monitoring and conflict resolution mechanisms ought to be different.

Extractive Reserve Rio Ouro Preto

With an area of 2,042 sq km, the Rio Ouro Preto Reserve (ERROP) is the smallest of the three reserves created in 1990. According to Ibama, there are 120 families living in the reserve (Ibama, 1999), but other sources quote figures as low as 75 families (Wawzyniak, 1994). In either case, the reserve inhabitants form a small group, a fact that in principle should be conducive to the development of common property. Rio Ouro Preto, however, is the only reserve that has had significant levels of deforestation – 3.71% (Ibama, 1999).

The failure of the Rio Ouro Preto Reserve to ensure the conservation of its forests lies on a combination of internal factors and external factors. In the 1980s, in the context of the MDB campaign one of the most publicised projects in Amazonia was Polonoroeste, located in the state of Rondônia (see Chapter 2). The state government was anxious to recover from the adverse publicity it had received on account of Polonoroeste, and to secure further World Bank loans for a new major project, PLANAFLORO (Rondônia Natural Resources Management); it thus decided to set up an extractive reserve and did so without consulting first either the inhabitants of the area or their representatives (Wawzyniak, 1994). The result of this approach is that the reserve's boundaries are grossly incorrect – privately owned areas without extractivists living in them were included in the reserve, and it is in these areas that all deforestation occurs (Irving and Millikan, 1997). There had been efforts to change the borders of the reserve, however, a perverse effect of the reserves having such a strong legal back up (see Chapter 4), is that it is extremely difficult to change their boundaries.

Besides the number of co-owners, there are two other important differences between the inhabitants of the Rio Ouro Preto and Chico

Mendes extractive reserves. One is the degree of collective action experience. In 1990, when Ibama created the Ouro Preto Reserve, rubber tappers in the area had hardly any experience of collective action (Wawzyniak, 1994). There were confrontations in the area between tappers and patrons because of the rent the latter charged for the rubber stands, and in the 1985 First National Meeting of Rubber Tappers in Brasilia there were tappers from Rondônia, mainly concerned with the pensions of former rubber soldiers (see Chapter 3). It was only in 1989, however, thus nearly 10 years later than in the ERCM, that the political organisation of the tappers in the areas began (Wawzyniak, 1994).

The common pool resource of the Rio Ouro Preto Reserve also differs substantially from that of the ERCM. The forests of the ERROP are yearly flooded. Rubber tappers have organised the distribution of their resources according to this fact: most of them hold the rubber trails in the flooded areas, and their agricultural plots in the drylands; during the summer, they work in the rubber trails, and during the winter they move to their agricultural plots. The latter, however, are sometimes located in other tappers' stands, which shows how the distribution of resources among the tappers varies between extractive reserves. During the winter, some families also move to the nearby cities (Wawzyniak, 1994), which increase their access to information, especially in comparison with those families of the Chico Mendes Reserve who sometimes are more than a year without talking with anyone outside their close set of neighbours.

As in the Alto Juruá Reserve, and contrary to what occurs in the Chico Mendes Reserve, tappers in the ERROP are grouped in small communities. In total, there are seven communities in the reserve. This suggests that it might be easy for them to develop robust institutional arrangements, and that a system of nested enterprises may be appropriate. Then again, it may be easier for the tappers to organise themselves according to the division of the reserve in flooded and dry areas. Only the resource users themselves, however, can identify what is the best arrangement, since they know with whom they feel sufficient affinity to cooperate and develop an institutional arrangement to manage their common resources. Ultimately, they know who they *want* to cooperate with.

Conclusion

This chapter has assessed the capacity of the Chico Mendes Extractive Reserve to ensure the conservation of the forest, and has provided an insight into the relationship between external and internal factors in the context of 'established' common property regimes.

The evidence suggests that the main reason why the performance of the ERCM in terms of promoting sustainable development has fallen short of expectations, is that although the reserve's institutional arrangements are largely based on those the rubber tappers had before its establishment, the reserve inhabitants do not see it as such. For the large majority of them, Ibama made the rules of the reserve, and hence the responsibility of monitoring compliance with the rules belongs to Ibama. Why do they think this although there were meetings to discuss both the continuation of the existing rules and the reasons for adding new ones? One reason is that although the tappers had a reasonably well-defined institution, they had not designed it themselves with any specific purpose, let alone for sharing a common resource. It had evolved over the years, and tappers wished to keep it as it was; as problems had never arisen, changing the rules or developing monitoring and enforcement mechanisms had been largely unnecessary. The concept of getting together and agreeing on a set of rules for ensuring a common good was thus largely alien to them, and with regard to the conservation of the forest many also thought it unnecessary.

Convinced that, devoid of a robust institution, the forests in the ERCM would promptly be destroyed, instead of providing only help and incentives for the forest dwellers to strengthen their institutions, external actors attempted to establish a robust regime. Both Ibama and the leadership endeavoured to include the reserve inhabitants in the process, but their strategy took insufficient account of the forest dwellers' characteristics. The leadership is formed by well-informed tappers, who have considerable experience in collective action initiatives. The other inhabitants of the reserve, however, have had hardly any experience of collective action, lack confidence to participate in discussions with external actors, and many are unaware of the problems facing the reserve. The establishment of the reserve – creation of an Utilisation Plan, establishment of Associations, and creation of a monitoring commission – represented too many new things on one go to a population with little experience of change and novelty. Besides, the process was too fast – all these institutions were to be established in less than a year, an insufficient time span for forest dwellers to perceive the need for them and to adapt them to their local conditions.

Two questions that arise, however, are: How long will it take the forest dwellers to strengthen their institutions? And, is there the risk that the forest is depleted before they develop mechanisms to prevent it? Ibama officials and the leadership believed that such a scenario was possible, for incentives to misuse the forest have increased, and, as tappers have insufficient experience in collective action, once they have perceived the need to manage the forest they may take too long to develop effective institutional arrangements. It can also happen that regimes develop on some estates but not on others, a situation that would jeopardise the entire reserve. Moreover, if the forest begins to suffer, tappers might lose all confidence in the reserve as an institution, as well as in any sort of communal action, which would further restrict the possibility that a robust regime would develop in the reserve.

The evidence suggests, however, that implementing a robust regime for the reserve was not the best strategy to prevent depletion of the forest. First, the main threat to the tappers' forests lied in their poverty rather than in the weakness of their regimes. The reserve inhabitants lack access to medical services, have insufficient schools for their children, and largely live a subsistence life. The more they see how people live in the cities, the more they perceive the shortcomings of living in the forest (although in fact many people in the cities live considerably worst than they do). A powerful incentive for tappers to sell their stands to loggers or to switch to unsustainable activities themselves is the possibility of using the money earned to move to the cities, where they believe they can have a better quality of life. Improving the living conditions of the reserve inhabitants would hence go a long way in diminishing incentives to break the established rules, whereas formalising the tappers' rules in the Utilisation Plan, setting up Associations, and training tappers to become monitors have achieved little.

Second, at the level of the rubber estates, there are incipient common property regimes with considerable potential to develop and hence ensure the protection of the forest. Providing that at the estate level tappers comply with the rules, the reserve as a whole is protected. Despite their shortcomings, the Unions could have continued to alone represent the rubber tappers, without the need for creating a new organisation. Setting up the co-management system of the reserve has hindered the forest dwellers' capacity to develop responsibility for the management of their resources and therefore jeopardised the sustainable use of resources in the long-term.

This is not to say that the co-management system of the reserve should not have conditioned the tappers' rights to their sustainable use of the resource. But, there is a difference between establishing rules with which commoners have to comply, and presenting such as rules as the result of the commoners' own decisions, which limits their interest and capacity in genuinely developing mechanisms to manage the forest.

This chapter has suggested that the best solution for the ERCM, and one that is worthwhile considering for other large reserves, is the development of a system of nested enterprises. The rubber tappers have incipient common property institutions at the level of the rubber estates, but these institutions alone cannot survive in the long-term. Once strengthened, however, the rubber tappers can harmonise their estate-regimes. It would be easier to encourage different communities to co-operate after they have gained experience in resource management through the management of their own estates. It is easier to establish co-operation between estate-groups than between all the tappers of the reserve also because as groups, the inhabitants of other estates are similar to them (equally dependent on the forest, practicing the same type of activity, same knowledge and appreciation of the forest[22]), whereas as individuals they are seen as different.

By comparison with the current system, which treats the reserve largely as a single and homogeneous unit, a system of nested enterprises would have at least two advantages. First, it would be grounded on existing institutions. The area that is now the reserve always had the same property rights institution (division of the forest in intertwined rubber stand and sharing of some common areas), but it never had a centralised organisation responsible for managing this institution as under the extractive reserve system. With the establishment of the ERCM, the property rights institutions of the area have been formalised, which gives more security and protection to the forest dwellers, but there has also been an attempt at centralising the management of the area under the umbrella of newly created organisations, such as the Associations. Second, a system of nested enterprises would allow for the differences between estates. In some respects the whole area is similar, but there are also important differences: on some estates, letting non-cooperative tappers participate in the use of common facilities serves as an incentive for further participation; by contrast, on estates where tappers have had negative experiences with free-riding, such a strategy may rapidly discourage them from participating in common endeavours[23].

In the same way that estates differ, so do extractive reserves. As exemplified by the Alto Juruá and Rio Ouro Preto Reserves, each reserve has its own jointness and exclusion conditions, their population is different, and they have different histories. Hence, their institutional arrangements need also to be different. Reserves share some common features: in all of them families hold their resources in a system of common property, and in most cases their institutions are weak and unable to cope with external threats. To recognise common property institutions as the extractive legislation has done is thus advantageous to all of them; to receive external support to strengthen their institutions and address the problems of poverty is also useful to all reserves. To specify how extractivists' institutions should be, however, is ineffective, for outsiders lack sufficient information about the resource, about the commoners and about how the latter use the resource. This, however, does not mean that extractivists and other commoners ought to be left to their own devices. The external context affects them, and it is beyond their power to change it or even to adapt to the changes the external context imposes on them. This and other reasons, makes imperative that commoners are supported, an issue discussed in more detail in the next chapter.

Notes

[1] To obtain a better overview of the reserve as a whole, I interviewed forest dwellers from different rubber estates. The main criteria for selecting rubber estates to visit were their accessibility and proximity to different cities, which were the two factors that key interviewees and other people in the area thought were the key determinants of the differences between estates. My own observations during the visits to these estates as well as the interviews with their inhabitants confirmed the fact that accessibility is a crucial factor distinguishing the different estates, as I argue later in this chapter. The five estates visited include three which are near the city of Xapuri, which is the one nearest to the state-capital of Acre, and which is where Chico Mendes was murdered, and two estates near Assis Brazil, which is considerably more inaccessible, it is one day away by car from Rio Branco, and the road is impassable during the winter season. Each estate has a different level of accessibility to the nearest city.

Rubber estates visited

Rubber estates	Distance to nearest city – Xapuri	Rubber estates	Distance to nearest city – Assis Brasil
São Pedro	1 day by boat	Guanabara	12 hours walk
Floresta	5 hours walk	Icuriã	5 hours walk
Dois Irmãos	5 hours walk		

[2] Some of the key informants were former tappers now working full time for one of the rubber tappers' organisations such as the rural workers' unions and who go regularly into the reserve. The other key informants were people who had never lived in the forest, but who worked in the reserve and so had a good knowledge of the area – e.g. academics who had done social research concerning the rubber tappers, people who had followed the process of formation of the ERCM, or technical staff working for development projects inside the reserve. The choice of key informants was based on snowball sampling and chance opportunities; 10 key informants were interviewed.
[3] This error was found out in 1991, upon the demarcation of the reserve boundaries carried out by IBAMA in co-operation with local teams and CNS representatives.
[4] For instance, if one of the daughters or sons gets married, the new family sometimes stays in the same stand but the young couple builds their own house and does not work in common with the parental family. The stand is divided and some rubber trails are given to the new couple. This could be explained by the fact that rubber tapping is an individual type of work. However, the new family also has its own agricultural plot. The son will help the father in his agricultural plot, and if the father is not too old he will help the son too, but each stand belongs to one family only.
[5] Although the boundaries had been demarcated in 1991, there had been several problems regarding the regularisation of the land tenure situation in the reserve because of the disputes over landed property rights in the area and the existence of land titles granted by different bodies. In addition, at the time of the creation of the reserve, IBAMA had no experience with expropriation procedures and there was little co-operation with other governmental organisations on the matter, such as INCRA (ELI, 1994; Irving and Millikan, 1997). The delays in the regularisation of the land tenure situation contributed to the continuation of conflicts over access to natural resources between rubber tappers and non-tappers who have land titles to areas in the reserve (Irving and Millikan, 1997). In the second half of 1992, however, IBAMA obtained legal rights to take possession of the lands in the reserve and was able to expel former owners.

[6] By October 1997, the co-management contract of the ERCM was not yet concluded (Irving and Millikan, 1997).
[7] The union of Xapuri is the representative of the areas of the reserve in the municipalities of Rio Branco and Xapuri; the union of Assis Brasil is the representative of the municipalities of Assis Brasil and Sena Madureira. To simplify the text, when saying the Assis Brasil or Xapuri area it is meant the areas under the jurisdiction of these unions, rather than the municipalities' areas.
[8] Interviewees considered that the representative of AMOREAB was not doing a bad job, but added that they would not have chosen him if given more time to decide because this man did not have the necessary experience and leadership characteristics they considered necessary for such a position. The organisers of the meeting, however, wanted to have a person with little experience in order to democratise the leadership of the reserve, which many believed was being kept in the hands of a few families of union members.
[9] According to a key interviewee, an additional reason why it had been decided that the unions should not be the representative organisations of the reserve was to avoid this separation between leadership and reserve inhabitants. The aim was that the reserve inhabitants themselves should participate in the management of the reserve and not former rubber tappers, like most of the lideranças. However, as shown before, although this objective was achieved regarding AMOREAB, the creation of the Association was not successful in terms of ensuring the forest dwellers' participation in the management of the reserve. In Xapuri, moreover, the separation between leadership and forest dwellers was reproduced in AMOREX. The Association representatives in Xapuri are formed by members of the union who live only part-time in the forest.
[10] Fiscalização da Reserva, 27. 'Cada seringueiro é um fiscal de sua colocação e das outras colocações, cabendo a ele não só zelar por sua colocação, como também observar para que as normas deste 'Plano de Utilização' estejam sendo cumpridas pelo conjunto dos moradores'. 28. 'Será constitutuída uma Comissão de Protecção da Reserva ligada directamente à Associação. Essa Comissão será composta por cinco membros eleitos em Assembléia Geral da Associação'. 29. 'O regimento da Comissão de Protecção da Reserva será elaborado pelo Conselho Deliberativo da Associação e aprovado em Assembleia Geral' (Ibama, 1995, p.27).
[11] The only rules that are not part of Brazilian law or the tappers' rules are those concerning hunting with dogs and hunting only for consumption purposes. These rules were deemed necessary because hunting with dogs scares the other animals away and there are now more opportunities for tappers to commercialise game.
[12] Although CNS has made alliances with the Indian tribes (see Chapter 3), the forest dwellers in general considered Indians as people different from them.
[13] Naturally, at the meetings organised in the reserve, tappers from different rubber estates meet. Yet, at none of the meetings attended by the researcher were there

attempts to articulate the incipient management regimes of the different rubber estates – the focus was more on merging all the estates into one big estate, the reserve. More recently, Ibama has tried to encourage the development of common property institutions on the rubber estates, but as it is discussed in the conclusion of this volume, instead of strengthening existing institutions they have mainly tried to implement new ones.

[14] Although as there is trading and inter-changing of rubber stands families do move in and out of a community.

[15] An increase in population could threaten the joint use of the resource. However, as is the case with other common property regimes (see Chapter 1), tappers can encourage migration and restrict utilisation of the forest by the members of the group if such an increase in population occurs. The development of new sustainable activities in the forest, which is one of the components of the RESEX sub-project (see Chapter 4), can also change the conditions of the CPR; for example, agroforestry may allow for a higher number of resource users than rubber tapping.

[16] It should be noted, however, that participation in communal work was not perceived only as a cost, there were tappers who saw it as an opportunity for socialisation, especially the younger ones. Nevertheless, once married and with more responsibilities, they could only afford to participate in communal works if this paid off.

[17] Most interviewees were men, and most leaders are men too. However, there are some exceptions, such as a former female rubber tapper, Marina Silva, who after being a leader on her estate, is now a senator for the state of Acre. As a general rule this book uses the term 'he' rather than 'he or she' to simplify the writing, as well as to avoid giving the impression that there is equal political participation of men and women.

[18] According to de Paula (1991), the presence of the middlemen also helped to curtail the dominance of the patrons. See Chapter 3.

[19] It should be noted, however, that some middlemen are tappers themselves. On rubber estate Icuriã, for instance, a middleman/tapper was one of the more active participants of the activities for the creation of the Association of Inhabitants of Assis Brasil, whilst he had never participated in co-operative Primavera.

[20] The forest branches of the co-operative were part of a decentralisation initiative to overcome the successive problems that the co-operative had faced; on account of which its results have fallen below expectations. See Hall, 1997b, p.117.

[21] Leaders living full-time in the city are treated here as part of the lideranças.

[22] Some rubber tappers interviewed, for example, mused on how nice it would be to live in the city; however, they usually added that for this they would have to be more qualified, otherwise living in urban areas would be considerably worse than living in the forest because they would not, for example, have access to clean water.

Once the decision to stay in the forest was made, tappers strongly believed that it was better to live in the forest than in areas that had been cleared.

[23] For example, on rubber estate Floresta a certain level of free-riding tended to increase participation in common works. By contrast, on rubber estate Guanabara even a small level of free-riding might discourage tappers to participate because of a previous experience with free-riders on the estate. In the 1970s, Padre Paulino encouraged the tappers to set up a co-operative. This arrangement, however, was short-lived because the tappers in charge of managing the co-operative used the common resources to their own advantage

6 The Interaction of Local, National, and International Factors in the Sustainability of Common Resources

Introduction

There are no simple and ready-made solutions for the conservation of shared natural resources. Privatisation, state control, and more recently community management have all been alternatively suggested as the best way of ensuring their conservation, but although on paper the case for any of these systems can be convincingly argued, in practice, situations are more complex, and choosing the appropriate property rights system is not in itself a guarantee that resources will be used sustainably. Extractive reserves are a case in point that both the adequacy of any property rights system and the sustainability of common resources also depend on the wider context and on how resource users interact with it.

The preceding chapter pointed out a series of problems with extractive reserves: the rubber tappers often fail to *manage* their common resources and, by and large, they show insufficient interest in the matter; many extractivists are abandoning the reserves, and the state interferes in the management of the reserves. Would these be reasons to argue that common property was not after all the best solution for the rubber tappers and to claim that extractive reserves have failed? Such a conclusion would ignore the following: first, the weakness of a common property institution is not in itself a justification for switching to another system such as private or state property; and second, what happens in the reserves is not solely determined by the features of the extractive reserve model, but also by how circumstances around the tappers develop.

The present chapter brings together the different parts of this study and discusses the interaction of internal and external factors in relation to the choice of property rights institutions, the development of common property regimes, and the capacity of these institutions to ensure the conservation of the forest. In so doing, this chapter shows why despite their weaknesses the reserves are nevertheless the best solution for the rubber tappers, and assesses the capacity of the theory on common property institutions to explain the sustainability of common resources.

Beyond private, state or common property: the diversity of property rights institutions

As was argued in Chapter 1, there are two different arguments regarding common resources. One is the 'tragedy of the commons' thesis, which postulates that unless there is no scarcity, individuals sharing a common-pool resource will deplete it, as they will ignore the full cost of their overusing the resource. Broadly speaking, scholars sharing this view can be divided into advocates for state control, who maintain that the difficulties of conserving a common-pool resource are similar to those of providing a public good and, therefore, private agents cannot secure the sustainable use of the resource, and advocates of private property, who consider that jointly used resources should be privatised because the owner would then carry the full cost of overusing the resource, and hence would only do so if the overall benefit was higher than the overall cost. The other line of argument is that even under conditions of scarcity, joint users of a common-pool resource can secure its conservation by developing a variety of property rights arrangements; consequently, depletion of jointly used resources only occurs if co-owners cannot secure exclusive rights to the resource, or if they cannot develop mechanisms to ensure their harmonious use of the common-pool.

Initially, the tappers exploited the forests under the control of a rubber baron or a patron, but from the 1920s onwards, *seringalistas* began to gradually abandon the rubber estates, and extractivists in eastern Acre were left to their own devices until the 1970s, when cattle ranchers arrived on their areas and tried to expel them. In the period between the patrons' departure and the cattle ranchers' arrival, the rubber tappers used the forests jointly and sustainably.

The preservation of the rainforest during the period in question has been considered to support the argument that joint users of a common-pool resource can, by themselves, secure the conservation of a shared resource. The rubber estates, however, were not conserved because tappers had developed robust institutions that ensured their using the forests sustainably. Between the 1920s and the 1970s, the common property regimes of the tappers of eastern Acre were very 'weak' (Ostrom, 1990). The boundaries of the common-pool and of the group of co-owners were well defined – a new comer, for example, could only establish himself on a rubber estate if the tappers in the area agreed with it – and there were rules for using the trees and other resources. These rules, however, had not been designed by the tappers to ensure the harmonious use of the common-pool, compliance with them was unmonitored, and the only enforcement mechanism tappers had was a small degree of social pressure. Institutionalised cooperation – for instance, clearing of forest paths – was on many estates non-existent. The common-pool resources of the rubber tappers, the rubber estates, consist of a set of intertwined 'private' rubber stands; at the time, boundaries among the stands were respected, and many resources were used in common, but the estates were not managed in common.

The conservation of the rubber estates before the arrival of the ranchers can be explained with reference to Hardin's (1968; 1991) analysis of the commons. According to Hardin, the 'tragedy of the commons' occurs only under conditions of scarcity. If there is abundance, individuals can jointly use a common-pool resource without depleting it. Before the government policies of the 1970s, which triggered a massive increase in the demand for land in the region (Chapter 2), this was indeed the case in Amazonia – resources were abundant. The rubber estates were largely inaccessible, and except for sporadically some outsides wanting to move into the area to tap rubber, established tappers were the only ones interested in using the estates' forests. For the extractivists alone, there was ample supply of natural resources. The use of each rubber stand was limited by the labour capacity of the household, and once the children were able to tap rubber themselves, there were enough rubber stands for them to move out without overusing the parental stand.

The conservation of the rubber estates' forests until the 1970s can also be explained using the arguments of the property rights school, which holds that a private owner will conserve the resource as long as the benefits of doing so are higher than the costs. Each rubber stand is private property in the sense that the owner of the stand has most of the rights to the *colocação*: the right to tap rubber, build a house, clear an agricultural plot, and sell the stand. Consequently, the stand owner carries both the benefits and (most of) the costs of destroying the rubber stand; if a family overuses the rubber trails, they have to find new trails to clear, and get a lower price for the existing stand. It can thus be postulated that rubber tappers used the forests sustainably because the benefits of doing so – maintaining their source of livelihood, selling the stand for a good price – were higher than the costs.

Notwithstanding the above arguments, the tappers' joint use of the forests also provides broad support for the theory on common property regimes. First, even in the absence of monitoring mechanisms, imposed by either the users themselves or the state, extracting as much as possible from a common-pool is not always the resource users' dominant strategy. The tappers had fewer incentives to deplete the resource than the cattle ranchers because for them the value of the stands lay in the standing forest, whereas for the ranchers the value of the holdings was in the cleared land – the existence of trees greatly diminished the commercial value of ranches. This indicates that it cannot be assumed that the dominant strategy for joint users of a common-pool resource will always be to extract as many units as possible; whether individuals try to do so depends on their economic and cultural context and on how natural resources are valued in that context.

Second, it is unnecessary to always have an external agent imposing rules of resource exploitation on commoners. Autonomous rubber tappers lived alone, free of patrons and other external agents, but nevertheless they respected the 'private' property rights to the rubber stands. Although resources were plentiful, the tappers could have attempted to occupy the neighbouring stands, which would have involved less work than clearing unoccupied spots. In the same way that the barons obtained control over large areas of the forest, individual tappers could have attempted to use their own rubber trails as much as possible and then take over those of their neighbours. There is no evidence, however, that they did so[1].

Third, on the rubber estates common property was a better option than both private property and state control. Privatisation was attempted but without success. This was partly because rubber estates are indivisible; the rubber trails are intertwined and water and other vital resources are unevenly distributed. Besides, the tappers' cultural and historical background made privatisation of the estates very difficult. Extractivists always had individual rights to the rubber trails, but fruits trees and other resources were freely accessible to all members of the community; the notion that use of these common resources would be barren to all but the private owner was a concept to which they found difficult to adapt. Private ownership of the entire estate was inadequate for many reasons. In the 1970s, private owners would not have secured the conservation of the rubber estates because at the time the private benefits of clearing the forest were considerably higher than the costs – the commercial value of the land was highest if cleared of trees and transformed into pasture. What is more, privatising entire estates made rubber tappers landless. Hypothetically, someone wanting to exploit rubber in the area could have bought the estates and employ the tappers to work in them. This in fact was attempted in the 1950s, but without the strong control of *aviamento* it was impossible to ensure regular production and profit (Hecht and Cockburn, 1989). State control of the entire area also presented important shortcomings. In the 1970s, and similarly to private owners, the state was unconcerned with the conservation of the forest. In the early 1990s, the protection of the Amazon rainforests is in the interest of at least Ibama, the federal environmental agency, but the state lacks the capacity to devise appropriate rules for the use of the forest, and to monitor and enforce them without the active cooperation of the forest dwellers.

The case of the extractive reserves provides a clear illustration of how diverse property rights institutions can be and this is why the insights of different theoretical frameworks may be applied. The property rights structure of the reserves involves elements of private, state and common property, and this distribution of rights between state, 'communities' and individuals is, for a number of reasons, more advantageous to the tappers than other alternative property rights institutions. The presence of each of these elements has a justification based on the history of the tappers, on the characteristics of the rubber estates, and on circumstantial factors related to the external context. The irregular distribution of resources in the forest and the tappers' use of the estates, which has traditionally involved individual work in the stands and common use of other areas, made both strictly private and communal[2] property institutions inadequate. Contextual socio-political factors, such as the struggle for land in the Amazon region, where small peasants are sometimes forced, through legalistic or violent means, to sell their plots, was also a motive for proposing state property of the land in the reserve. The fact that the reserves were established amid international pressure for the conservation of the Amazon rainforest, and in the context of the government's environmental policy were a further motive behind state property: through state property, it was argued, reserves could be more easily justified as environmental units. Specific cultural factors also played a role in determining the property rights features of the reserves. For the majority of the tappers, 'private ownership' of the stands does not include the right to sell land and the right to clear the forest cover (in the same way that private ownership of a city house seldom includes the right to demolish it). Other users of common-pool resources, such as agricultural colonists whose aim in migrating to Amazonia was the acquisition of a private plot of land for agriculture, would have opposed a property rights system that excluded the right to sell the plot and clear the land. Property rights institutions can thus take on a number of different characteristics, depending on the features of the resource, its direct users, and the socio-economic, political and historical context where resource and resource users are embedded.

The development of common property institutions

Chapter 1 suggested several factors that can influence the development of common property institutions, some pertaining to the characteristics of the resource and its users, and others to the external context. Whether commoners set up a common property arrangement regime is contingent on the potential of the common-pool resource to be held as common property; for instance, on how difficult it is to prevent non-owners from using the resource. Resource users must also perceive the need to establish a resource management regime, and for this to happen the resource must be relatively scarce and commoners must possess sufficient ecological knowledge. The commoners' dependency on the resource, and the size and homogeneity of the group they form, are also factors to consider. The state and private external agents, the socio-political context, and the legal setting can affect the joint use of a resource by triggering changes in the circumstances, and by influencing the commoners' capacity to address changes induced by either internal or external developments.

As is the case with all common-pool resources, exclusion from the tappers' forests presents some difficulties, which result from the large size of the resource, if we consider the latter as all the forested lands with extractivists living there. It is possible, however, to define a particular area such as a rubber estate, a village (in the case of Rio Cajarí Reserve) or an extractive reserve, and prevent outsiders from entering. If the area defined is sufficiently large to ensure the ecological processes that maintain the forest, such as pollination, tappers can conserve their resources and be only marginally affected by the use of the surrounding forests. The jointness conditions of the rainforest are also conducive to a common property regime because provided that the trees are not cut down, several individuals can use an area of forest without imposing costs on fellow users.

Several observations suggest that rubber tappers perceiving the need to develop a regime was one of the most important conditions for the development of common property institutions on the former rubber estates. Before the arrival of the ranchers, tappers had never attempted to establish strong exclusion mechanisms, but once the need to secure boundaries became clear, because if shared with ranchers the forests were scarce, extractivists engaged in effective collective action to obtain exclusive rights to their resources. Likewise, tappers never attempted to develop mechanisms to ensure their own harmonious use of the common-pool resource; by and large they were unaware of the need to do so. Common property arrangements on the rubber estates of the ERCM were also related to the extractivists perceiving the need for them. The highest records of participation were observed in initiatives from which many tappers could benefit; clearing forest paths in common takes place only on estates where the main users of the paths are tappers rather than middlemen; and whereas the use of common facilities is monitored, the use of the rubber stands is not – tappers consider important to prevent individual tappers from taking over what belongs to all, but they are generally unconcerned with the implications that the misuse of one stand may eventually have on the entire common-pool resource.

The strong dependency of the rubber tappers on the forest was also a contributing factor to the development of common property on the rubber estates. Outside the forest, rubber tappers have few means of subsistence, partly because they lack the necessary skills to find work in the cities. A direct relationship between tappers realising their dependency on the forest and engaging in collective action to protect their common-pool resource was observed in at least two instances. First, resistance against the ranchers started after tappers had experienced poverty in the cities and become aware that despite the shortcomings of living in the forest, their life there was better than in the nearby towns. Second, tappers began demanding the right to stay on the rubber estates after realising that even if ranchers compensated them for the loss of their rubber stands they were nonetheless unable to make a living in either the cities or the rural areas.

In the literature on common property, there is a debate on how important the size and homogeneity of the group of users is for the development of robust institutions. According to Singleton and Taylor (1992) only a group that is small (around 300 users), whose members are homogenous, have direct and multiplex relations beyond their collective action problem, and are mutually vulnerable, *can* endogenously solve their common-pool resource problems. Conversely, Ostrom (1992) argues that, first, a group meeting these characteristics *may not be able* to secure the conservation of their common-pool resource if the external context is obtrusive, for instance, if the state fails to recognise the users' common property regime; and second, if all other factors are equal, small groups are more likely to develop common property institutions than large groups, but large groups can also establish such institutions.

To develop a robust common property institution is necessary to address two different problems: first, how to exclude outsiders, and second, how to harmonise the co-owners' use of the resource. The evidence from the Chico Mendes Reserve suggests that to solve the first problem it is not sufficient that commoners form communities as defined by Singleton and Taylor (1992). Before the arrival of the ranchers, the inhabitants of each rubber estate formed small and relatively homogeneous groups, and to a higher or lesser extent had direct and multiplex relationships. Had they had more regular contact with each other, could the tappers have prevented outsiders from taking over their resources? This is a moot point that can only be speculated upon. The research conducted, however, suggests that even if tappers formed more tightly knit communities, without external help they would have failed to obtain secure rights to their forests. The rubber tappers' capacity to resist eviction was restricted by their being unaware that legally they had usufruct rights to their stands, and hence in theory at least could count on the support of the courts; that outsiders provided them with this information was vital for strengthening their resistance. External agents were also important because they supplied the extractivists with organisational know-how. Another problem was that the ranchers were more powerful than the tappers. They were backed up by the government's interest in cattle production and had support from official organisations such as the police, whereas the tappers had to cope with the government's dismissal of the rubber trade and disregard for their property rights institutions. It was through the help of external agents that the tappers made contacts with powerful allies, especially international environmental NGOs, and obtained sufficient political leverage to ensure protection for their rights. If external agents are more powerful than the commoners, even if the latter are small and homogeneous groups they are very likely to need external help to secure the conservation of their common resources. Institutional theory often notes that communities are in general unable to deal with a controller state; a less mentioned fact, however, is that communities can also be unable to deal with powerful private agents.

Do users of a common-pool also need external help to harmonise their own use of the resource if they form small groups as defined above? The examination of the Chico Mendes Reserve suggests that the factor distinguishing rubber estate communities that have developed common property from those that have failed to do so is not the number and homogeneity of their members but the extent of external help they have received. The poorest communities tend to be those that have received hardly any help; tappers from these communities are more likely to bend to outside pressure for destroying the resource system (e.g. by selling timber), than tappers in more 'well off' communities. If left to their own devices, small groups may take a long time to perceive the advantages of managing their common resources. Also, they may have insufficient financial resources for any initial investment necessary for a common endeavour, such as buying animals to set up a co-operative, or taking days off work to clear a forest path. Also, their knowledge of the resource may be good, as they have been using it for a long time, but they may be unaware of the full potential of the resource. The incentive provided by outsiders pointing out what they can achieve together, and supplying the basic material means can be crucial in promoting the development of common property regimes.

Regarding the ability of large groups to develop common property, it is worth noting first that large groups are sometimes *necessary* for resource management institutions to develop. For the stand-offs in Acre to be successful it was essential to count on the participation of tappers from many different estates; a small group of extractivists would have been powerless against the ranchers and their gunmen. Another reason why small groups are sometimes unable to establish robust regimes is that a common-pool resource may be too large for them to have total control over it: a small group of tappers can set up a rules and monitoring mechanisms to ensure the conservation of the rubber estate, but if all other rubber estates around them are being destroyed they will be unable to preserve their own patch of forest. But can large groups develop common property or when small groups cannot do it by themselves a 'tragedy of the commons' is bound to ensue? The evidence from the Chico Mendes Reserve is inconclusive. On the one hand, the lack of a robust regime can be pinned down to the reserve having such a large number of inhabitants, since at the level of the rubber estates tappers have been able to develop common property arrangements and, to a higher or lesser extent, develop an identity as a group. The only difference between the reserve and the rubber estates is that in the latter tappers form smaller groups. This, however, is an important difference, for in a small group the need to cooperate is more evident than in a large group, and as individuals meet more regularly, they can more easily develop a group identity. On the other hand, it can be argued that the reason why the reserve inhabitants have yet to develop a common property institution for the entire reserve is that they fail to see the need for it. It is also plausible that they have failed to establish an institution to manage the reserve because they think that the state is responsible for doing so.

This study postulated that the development of common property regimes depends on external as well as on internal factors. The case of the extractive reserves provides ample evidence on this matter, since the development of these institutions cannot be explained solely though the characteristics of the rubber estates, the tappers, and their informal institutions. The triggers that led to collective action – abandonment of the rubber estates by the barons, and arrival of cattle ranchers – were externally produced, and likewise the capacity of the tappers to deal with the ensuing changes was influenced by many external factors such as government policies, the legal setting, help from external actors, and the socio-political and economic context at the national and international levels.

The Brazilian government has played a significant role in the development of the tappers' institutions. During World War II, the establishment of a contract between patrons and rubber tappers enhanced the capacity of the latter to resist complete domination by their employers. Later, between the late 1940s and the 1970s, the state was 'indifferent' with regard to the rubber tappers: it neither hindered their institutions by attempting to manage the rubber estates nor supported them by backing up the extractivists' rights to the forest. In this context, and without the strong dominance of the rubber barons and the *aviamento* system, rubber tappers were able to develop their own arrangements to continue using the forest. In the 1970s, however, the government policies provoked a radical change in the tappers' circumstances, and this seriously threatened the capacity of their institutions to secure the conservation of the forests; rather than being a fortuitous event, the arrival of outsiders on the rubber estates was the direct result of new government policies for the region. The state also contributed to the depletion of the tappers' forests because it failed to protect their usufruct rights against the newcomers and, moreover, it promoted economic activities that require the removal of the forest cover. The destruction of the forests in the 1970s and early 1980s can thus not be explained by internal factors only; the factors that led to the incipient destruction of the tappers' forests – opening of roads, arrival of outsiders, and lack of secure property rights – were all related to the federal government policies.

The legal setting also influenced the tappers' capacity to ensure the conservation of their resources. When the extractivists began their struggle to protect the boundaries of their forests against outsiders, the existence of a legal item recognising usufruct rights to resource users helped them to resist expulsion by the cattle ranchers. Conversely, the absence of any legal item recognising the existence of common property rights rendered the tappers' efforts to formalise their institutions particularly difficult. In the late 1980s, the enactment of a new constitution made the tappers' legal setting more facilitative. In relation to this, it ought to be noted that not only legal items directly relating to common property can be helpful for commoners; the stipulations of the new legislation regarding environmental protection also helped the extractivists to obtain a stronger legal claim, the extractive reserves, to secure their exclusive rights.

The process that led to the establishment of the reserves was, from the beginning, embedded in developments in the socio-political and economic context in the national and international arenas. In the last decades of the 19th century, worldwide demand for rubber was what attracted the tappers to Amazonia. Later, international political developments, namely World War II, brought back to the rubber estates the former patrons and *seringalistas*, temporarily reversing the process of growing autonomy the tappers had experienced since the end of the rubber boom. In 1985, the transition to democracy in Brazil, the positioning of land reform on the political agenda, and the fact that the landless movement and other grassroots organisations were voicing their demands, provided the right setting for the tappers to organise a national meeting. Socio-political changes in the international context, specially the growing interest of public opinion with environmental issues and deforestation, also played a significant part during the second half of the 1980s. The NGOs' campaign against the MDBs was influenced by the political situation in the industrialised countries, and in this way changes of governments in the North, for example, influenced the tappers' struggle for exclusive rights to their resources. The creation of extractive reserves was, likewise, an initiative firmly entrenched in the national and global developments of the early 1990s. The changes that were occurring in Brazil, the international pressure on the Brazilian government to conserve the Amazon rainforest, and the upcoming United Nations Conference on Environment and Development created a supportive political and legal environment for the tappers to obtain an answer to their demands.

General trends in society had an effect on the tappers' capacity to conserve their common resources, too. The extractivists benefited from new ideas about the role that local communities can play in environmental conservation. Widespread acceptance of the view that common property institutions can ensure the conservation of natural resources, for example, made obtaining external support for the extractivists' institutions possible. In the past, propositions like the extractive reserve would have been ignored, for the accepted wisdom was that all jointly used resources were potential 'tragedies of the commons' and that besides local communities were incapable of protecting their own resources. Societal trends on issues not directly related to common property institutions were also important. For instance, in the 1970s the consequences of deforestation were virtually unknown to governments and public opinion; the Brazilian government policies in the 1970s and early 1980s, and the support MDBs gave to various environmentally and socially destructive projects in Amazonia were thus rooted in the overall understanding of natural resources at the time. Likewise, support for the tappers' regimes in the 1990s, through the Pilot Programme and other initiatives, is related to the changes that had occurred in society regarding activities such as extractivism, which instead of being considered 'backward' are now seen as 'sustainable'.

Finally, it should be noted that as the external context influenced the tappers' institutions, the tappers themselves also had a bearing on the external context. The tappers' confrontations with the cattle ranchers helped to publicise the negative impacts of deforestation in the industrialised countries. And more important still, the tappers' struggle determined an important change in the legislative framework: the enactment of the decrees on Extractivist Settlement Projects and Extractive Reserves made the legal setting for Brazilian commoners significantly better.

Ensuring the sustainable use of the forest

Although common property institutions can ensure the conservation of jointly used resources, they not always do so. Chapter 1 suggested that the conservation of a commons is contingent on the robustness of the common property institution, which depends on whether the regime has clearly defined boundary and harmonisation rules, along with monitoring, enforcement and conflict resolution mechanisms. The external context influences all the characteristics of a regime and their effectiveness. Judged by the theoretical framework, the Chico Mendes Reserve is a weak common property institution and its inhabitants lack sufficient autonomy to manage their resources (see Chapter 5). This study has shown, however, that the conservation of the forests in the reserve depends not only on strengthening the institution and on leaving more autonomy to the tappers. External help to encourage the forest dwellers to develop robust regimes and to address the issue of poverty are equally important requirements.

Boundaries must be well defined so that commoners know what they are managing and with whom they should cooperate. The Chico Mendes

Reserve has clearly defined boundaries and the inhabitants of the reserve have exclusive rights to their resources, recognised by the law and protected by the state. The boundaries are adequate to the physical characteristics of the forest: the area of the reserve is sufficiently large to guarantee the necessary ecological processes. The problem, however, is that unless they live nearby the reserve inhabitants are unaware of where the boundaries are. They have little interest in cooperating with the fellow tappers outside their rubber estates, and, besides, they are ignorant of whom the reserve inhabitants are.

An additional but lesser difficulty regards the external context. The reserve inhabitants are influenced by what occurs outside the reserve boundaries; if a predatory mode of production takes place next to the borders, this can increase the commoners' incentives to misuse their resources. Hence, land policies in the areas surrounding the reserve, such as INCRA's plans to set up agricultural settlements (Irving and Millikan, 1997), can strain the capacity of the tappers' institutions to prevent destruction of the forest. This problem has already occurred in other reserves.

Harmonisation rules must be locale-specific, clear, easy to enforce, and specifically designed to protect the common-pool resource. Commoners must define them and be able to change them if necessary. The rules of the ERCM, stated in the Utilisation Plan, are specifically aimed at the conservation of the natural resources in the area, and in general are clear and appropriate to local conditions. But they were not established by the forest dwellers, who besides are often unaware of their environmental purpose. This weakens the effectiveness of the rules first, because the tappers fail to feel responsible for them and their enforcement, and second, given a change in the circumstances, the forest dwellers will be unable to change the rules to adapt to the change. The rule proscribing the selling of land illustrates this problem. One of the reasons for this ban is that it helps protecting the common-pool resource. Ranchers and land speculators are only interested in buying land if they can later sell it, which they cannot do if they illegally buy land from the reserves; banning the selling of land thus reduces the tappers' incentives to sell their stands to individuals who would disrespect the environmental rules of the reserve. Since the establishment of the reserves, however, circumstances have changed. The ERCM as well as the Rio Ouro Preto Reserve are now mainly threaten by loggers, who are interested in taking trees out rather than in commercialising land. Tappers, however, have failed to adapt their rules to this new threat.

Monitoring and enforcement mechanisms are important because they fulfil the double task of ensuring compliance with the rules and reducing the co-users' incentives to free ride; everybody knows that the other members of the group cannot free ride on their efforts to secure the 'common good', and that if they attempt to free ride the possibility of being caught is high. Social pressure is a useful enforcement mechanism if the resource users form a small group and are checked by shame, but if the group is large (as in the case of the reserve), or the incentives to free ride are particularly strong, social pressure will be ineffective. Monitoring and enforcement must be the responsibility of the co-owners, who have a direct interest in the conservation of the commons.

According to the utilisation plan, the reserve inhabitants (together with Ibama) are responsible for monitoring the use of the reserve and enforcing the rules through various mechanisms. The formal structure of the reserve, however, is not mirrored in the actual institutional arrangements of the forest dwellers. The sole enforcement they engage in is social pressure, and this only in some areas. Besides, by and large tappers prefer Ibama to monitor compliance with the rules, especially with regard to the action of fellow tappers; consequently, environmental monitors have had some success preventing logging by outsiders (Hall, 1997b), but little tackling infringement of the rules by fellow tappers (Irving and Millikan, 1997).

The tappers' lack of interest in monitoring can put the conservation of the natural resources in the reserve at jeopardy. Whereas hitherto there have been only isolated cases of tappers breaking the rules, increased incentives and opportunities to do so can expand the level of defection. Both incentives and opportunities are contingent on the external context: the tappers' main incentive to switch to other activities is the little return they obtain from selling rubber, which is in turn related to the world demand for rubber and to the government policies for the sector; opportunities to switch to destructive activities such as logging also depends on the demand for tropical wood and on loggers attempting to buy timber from the reserve. A robust monitoring system can help prevent these problems threatening the reserve, but this requires that tappers take responsibility for their resources. The capacity of state agents to monitor compliance with rules is limited: their knowledge of the area is considerably lower than that of the tappers, and so any extractivist could easily deceive them; having external agents monitoring the reserve without the full support of the tapper community is also very costly, and thus economically unsustainable; and poorly paid state agents can be more easily bribed than the reserve inhabitants, who if they did, they would be accepting bribes from someone free-riding on their own efforts to conserve the forest. Tappers thus could enforce the rules more effectively. As they dislike checking on their neighbours, they could hire external monitors, but they would then be responsible for supervising the work of the monitors. However they choose to do it, tappers must first realise that the conservation of their stands depends on the conservation of all stands in the reserve and that external agents cannot ensure the conservation of the reserve without their participation.

There is no generally recognised authority or mechanism for solving conflicts in the reserve. The Utilisation Plan states that the Associations of Inhabitants of the Reserve are responsible for addressing the problems arising from lack of compliance with the rules, but if faced with a conflict with a neighbour rubber tappers seek help from the rural workers' unions and Ibama as well as from the associations. If these various organisations happen to disagree on the solution for a conflict, the situation may reach a standstill. Even if they reach an agreement, one of the conflicting parties is likely to feel unfairly treated if the organisation that proposed the solution to the problem is not the same that represents them; for instance, the Association may have proposed the solution, but the Union may be the organisation representing them. Tappers try also to solve conflicts between themselves, but if the conflict is particularly serious informal solutions are unlikely to be effective, and can be easily undermined by decisions from external or semi-external organisations like tribunals, the Associations or the Rural Workers' Unions.

External help and autonomy

External help is fundamental for ensuring the conservation of the forests in the reserves. Chapter 1 argued that external help, however, should not restrict the commoners' autonomy. That resource users have sufficient autonomy to manage their resources is as important for weak regimes as it is for robust regimes. If the regime is robust, it means that commoners know how to manage their resources, and interference by the state is unnecessary and, most probably, detrimental to the sustainability of the common-pool. If the regime is weak, as the tappers' regimes are, commoners will eventually need to strengthen it, and for this they must alter the rules and develop new ones; if outsiders do this for them, the rules can be inappropriate for the local conditions, and, moreover, commoners will refuse to take responsibility for them, which greatly weakens the institution. This is what seems to be occurring in the Chico Mendes Reserve.

The case of the ERCM illustrates how outsiders can restrict commoners' autonomy, even when they have the opposite objective. The aim of the state, the lideranças, and the RESEX sub-project is that the reserve inhabitants should manage their own resources. As the tappers' community institutions are very weak, however, they feared that the reserve inhabitants could deplete their own resources, especially considering that circumstances had changed since the 1960s, when there were hardly any incentives to misuse the forest. It was believed that if left to their own devices tappers would be incapable of expediently strengthening their institutions, for they form a large group, live scattered in the forest instead of forming tightly knit communities, and have little experience of managing resources in common. Hence, to ensure the conservation of the forests in the reserve, the state and rubber tappers' organisations formalised the forest dwellers' rules in the Utilisation Plan, created monitoring mechanisms, and encouraged the reserve inhabitants to form Associations and local groups (*núcleos de base*) to manage the reserve.

Yet, these initiatives have hindered the reserve inhabitants' autonomy to manage their own resources. Part of the problem is that participation of the leadership in the design of the reserves' institutional arrangements was treated as equivalent to participation by the forest dwellers. Most members of the leadership, however, have been living out of the forest for many years; they know the basic features of the rubber tappers' institutions, but are naturally ignorant of the specificities and problems of each rubber estate. Also, they perceive the established institutions in a different way from the forest dwellers, because their daily functioning and implications fails to affect them directly. The leadership have a different understanding of the common-pool resource: for them the common resource is the reserve, whereas for the forest dwellers it is the rubber estate, and often even only part of the *seringal*. And, the *lideranças'* proposals sometimes incorporate political ideals and objectives that are at variance with the characteristics of the reserve's inhabitants; the leadership are more community oriented than the forest dwellers, who in general are quite individualistic in their views.

As the forest dwellers did not devise the formal institutions of the reserve, some of the latter are inadequate to the local conditions. In the Chico Mendes Reserve, the creation of three Associations to manage the reserve is a case in point. The reserve inhabitants already had representative organisations, the rural workers unions, and from their point of view, having another representative organisation whose sole responsibility was the management of the reserve (the unions represent also the interests of rural workers outside the reserve) was unnecessary. The creation of organisations to substitute the rural unions may undermine the latter without taking their place.

For outsiders to develop an institution rooted in local conditions is extremely difficult. One year after the establishment of the Association, CNPT agents noted that participation in the management of the reserve was still scant, and that at the level of the estates there was more potential for communities to develop; consequently they decided to set up local groups on the estates (*núcleos de base*) (Irving and Millikan, 1997). CNPT's decision was grounded on local conditions; at the level of the rubber estate, there is potential for the development of robust institutions for there are incipient regimes, communities are small and formed by relatively homogeneous members who have direct and multiplex relations – although these factors cannot alone determine the development of a robust regime, they are generally conducive to it. Yet in 1997, two years after their establishment, an independent review observed that very few of these newly created communities was operational, people failed to meet regularly, and were uninformed of the purpose of these newly created organisations (Irving and Millikan, 1997). According to Ibama (1999), by 1999, these groups were participating in many communal activities, but it remains unclear whether these activities are individual works initiated by Ibama agents or the local groups have developed institutions, which require a regular and established set of procedures. The problem seems to have been that the *núcleos de base* were not grounded on the existing micro-communities of the estates, such as religious communities and small cooperatives, and that CNPT failed to discuss the proposal with the other representatives of the forest dwellers, such as the rural unions, which also have groups established inside the reserve.

Similar type of problems may be occurring in the other reserves. The association in the Alto Juruá reserve was created before the establishment of the reserve, and appear to be the genuine representatives of the extractivists, but it cannot be ruled out that other aspects of the formal institutional arrangement of the reserve at variance with the actual institutional arrangements of the forest dwellers. In both the Alto Juruá and the Rio Ouro Preto Reserves the forest dwellers have less experience of collective action than those in the Chico Mendes Reserve, and the Rio Ouro Preto reserve is one of the most threatened reserves; the lack of genuine community organisations and the corresponding motivation for external actors to implement mechanisms to prevent the depletion of the forest are thus likely to have created situations similar to those of the ERCM.

By attempting to *implement* common property regimes there is the risk that tappers rely on outsiders to manage their resources. Tappers perceive Ibama as the new owner of the reserve and therefore as the entity responsible for managing the entire area. Developing a robust regime requires considerable effort: commoners must identify their needs with respect to the resource, and discuss among themselves what to do, which can create disagreements and problems with neighbours. The actual management of the forest also requires time, which they have to deduct from their individual work in the rubber stands. Hence, if commoners believe that an external agency can provide the 'common good' they are likely to rely on this providing that it does not interfere with their individual use of the rubber stands, which is their main concern. Why should they lose workdays strengthening their monitoring devices and devising new ones if they believe others – environmental monitors from other rubber estates and Ibama officials – can do it for them? The fact that, at the time of the fieldwork, the formal institutions were very recent may partly explain the tappers' lack of interest in the Associations, the rules, and the monitoring. Yet by 1997, when the last independent review of the reserve was conducted, the tappers still considered the rules as Ibama's rules, and their participation in monitoring was limited (Irving and Millikan, 1997).

The two assumptions underlying Ibama's and the leadership's strategy – that there was a risk that the reserve inhabitants would overuse the forest because their community institutions were weak, and that if left to their own devices they would be unable to expediently strengthen them – are only partially correct. In the short term, the development of a robust regime for the entire reserve was not the most pressing requirement for ensuring forest conservation. The reserve can be conserved if the resources are conserved at the level of the rubber estates, and, in general, on the estates tappers have mechanisms to ensure to a higher or lower extent compliance with the rules. What is more, the risk of destruction of the reserve lay with the tappers' poverty rather than on the weakness of their institutions, which even if robust would hardly be able to deal with issues such as lack of access to doctors and schools, and the low price of rubber. With regard to the forests dwellers' capacity to strengthen their regimes, two points must be made. First, even if incapable of expediently developing a robust regime for the entire reserve, implementing a common property regime on the commoners' behalf is more likely to retard rather than advance the process. As argued above, the new institutions may disrupt the existing ones without taking their place, and forest dwellers may choose to rely on external actors for the common good. Second, although it is true that if left to their own devices tappers would find it difficult to develop a regime for the entire reserve, supporting commoners in their efforts to set up institutions is different from supplying them with almost ready-made institutions.

Poverty and the conservation of the forest

Chapter 1 argued that external actors, including the state, can support commoners by setting up arenas for discussion and conflict resolution, providing technical and expert information on the characteristics of the resource, capacity building activities and provision of small scale infrastructure. All this help is welcome in the case of the rubber tappers, who left alone, would find it difficult to develop robust regimes, for besides forming a large group with hardly any experience of collective action, they have little access to information. The rubber tappers, however, also require help to improve their social and economic conditions – addressing the problem of poverty in the reserves is important for humanitarian reasons and also because the conservation of the forests depends on it.

The four main threats to the conservation of the reserve's forests are tappers switching to predatory logging, cattle ranching and other activities which are ecologically unsustainable, selling their usufruct rights to non-extractivists, and abandoning their stands. In the Chico Mendes reserve, it is mainly 'well-off' tappers who rear cattle, and their neighbours realising that this activity can potentially jeopardise their living would go a long way to stop it. Logging, by contrast, tends to occur in remote areas, where tappers are very poor. In these areas, the tappers' poverty is like a vicious circle: they are poor because as the cities are very far, they depend exclusively on the middlemen for buying consumer goods and commercialising their produce; traders tend to charge high prices for what they sell, and to buy the tappers' produce cheaply, in this way maintaining the tappers in permanent poverty; the poorer they are, the less likely is that they can afford to take days off work to sell the rubber for a better price in the cities, let alone to buy animals to take their produce out of the forest; as they rarely leave their estates, they have little access to information on ways to improve their lives. An opportunity to earn what seems like a large sum buy selling timber is thus taken, in the hope that with the money obtained they can make a living outside the forest.

The development of a robust regime in remote areas could prevent tappers from selling logs: enforcement mechanisms could be high enough so that neither tappers nor loggers would be willing to risk engaging in forest clearing. Yet it could not stop tappers from simply abandoning the forest in search of a better life elsewhere. Abandoning the forests is in fact one of the main problems of the reserves: whereas in 1994 there were 1,838 families living in the ERCM (Feitosa, 1995), in 1999 there were only 1,465 (Ibama, 1999); in the Alto Juruá reserve there were 865 families living there in 1994 (Feitosa, 1995b), and only 665 in 1999 (Ibama, 1999). Once tappers abandon the reserve, it becomes easier for outsiders to enter through deserted areas and monitoring becomes more costly and difficult. Addressing poverty would help diminish the tappers' incentives to both undertake predatory practices and abandon their stands; in so doing, it would also facilitate the functioning of the common property regimes, since as incentives to defect diminish, rules enforcement becomes easier.

The improvement of the tappers' economic conditions is partly related to the development of robust regimes to manage common facilities and co-operatives to commercialise their produce. For this they need several forms of external help, such as information on how to set up cooperatives and financial support for the initial investment. They also need external help, however, to improve their access to education and health. Through collective action initiatives local communities can obtain better facilities, but external support is fundamental to train teachers for the schools, obtain medicines for health posts, and improve their access to doctors and hospitals.

G7 Pilot Programme to Conserve the Brazilian Rainforests

The main source of aid of the Chico Mendes Reserve is the RESEX sub-project of the G7 Pilot Programme for the Conservation of the Brazilian Rainforests (see Chapter 4), which also supports the Alto Juruá and Rio Ouro Preto Reserves[3]. The PP-G7 provides valuable support to the reserves, but it also tends to hinder the forest dwellers' autonomy.

RESEX includes initiatives to both strengthen the institutional arrangements of the rubber tappers and alleviate their poverty. The sub-project provides help to regularise the formal property rights structure of the reserve and delimitate the borders of the reserve; in so doing it strengthens the tappers' exclusive rights to their resources. RESEX supports as well the development of community organisation: it promotes meetings among the inhabitants of the reserves, supplies material infrastructure for the communities, and provides training in community organisation related areas, such as planning, administration, and accounting. With regard to poverty alleviation, the PP-G7 finances education and health projects, as well as the development of alternative technologies to improve the productive activities of the reserve inhabitants.

Like CNPT and the leadership, RESEX's aim is that tappers participate fully in the management of their reserves. To this end, the sub-project includes a component specifically aimed at promoting a participative method for managing and administering the project. Yet some of the help RESEX provides hinders the autonomy of the rubber tappers and in so doing it weakens their capacity to eventually manage their own resources. One of these initiatives was already discussed: the creation of the Associations in the ERCM. Another one concerns the implementation of alien monitoring mechanisms. On some rubber estates, tappers developed incipient monitoring mechanisms, such as locating common resources at a crossroads, where community members frequently pass, or in the stand of a trusted member of the community. But in the context of the RESEX sub-project common facilities are located in empty stands and monitoring is done by environmental monitors, a mechanism that involves more work and whose effectiveness is doubtful, since as mentioned earlier rubber tappers dislike the idea of checking on each other's actions.

A third initiative of the PP-G7 subproject that is inadequate to local conditions is to separate representation and management of the reserve. This division of responsibilities might be necessary in the future, but not at the time of implementation of the project. Communities in the reserve are still at an early stage of development, on most rubber estates, tappers are only just now being acquainted with what the reserve is, and the creation of two administrative levels can make things more difficult for them to understand and become familiar with. The separation of these two functions can also increase the existing gap between leaders, who mostly live outside the forest, and forest dwellers. With the creation of two administrative levels, the bureaucracy of the reserve augment, and this in turn results in the official representatives having to spend most of their time travelling and attending meetings in the towns, leaving them little time to carry out their productive activities in the forest. This means that they spend less time in the forest, their contact with their neighbours is reduced to the discussion of reserve matters, and their interests naturally become different. And, finally, the separation of functions increases the bureaucracy and hierarchy of projects, for it becomes necessary for the forest dwellers to deal first with the appointed person responsible for managing the reserve, then with the representative of the reserve, and in the end with the external organisations responsible for the ERCM.

Developments in the wider context

The reserves' inhabitants are also affected by developments in the wider socio-political and economic context, although not to the same extent as they were before their rights were legally recognised. When the extractivists had only informal rights to their stands, outsiders could easily take over their lands; with the creation of the extractive reserve, however, the tappers' exclusive rights are firmly entrenched in the law, and the arrival of outsiders in there area would not have the same devastating effect the arrival of the ranchers in the 1970s had. Furthermore, the existence of CNS, in spite of its shortcomings (Hall, 1997b), also enhances the reserves inhabitants' capacity to address changes in the wider socio-political context, as this organisation defends the tappers' interests in the wider political sphere.

The RESEX sub-project, changes in the government policies for rubber, and the development of new markets for non-timber forest products can illustrate the influence that the wider context has on the reserves' inhabitants. The inclusion of the RESEX sub-project in the Pilot Programme, and the delay in its implementation were the result of developments in the Brazilian and international political arena; whereas the anticipation of UNCED advanced the sub-project, subsequent developments such as the government change in Brazil in 1992, and the reduced interest of the G7 in deforestation after the conference, contributed to the near stagnation of the project and as a consequence to delay the support tappers should receive from the G7.

The national policy for the support of rubber production has been gradually withdrawn since the 1980s (FoE/GTA, 1997). To secure a market for domestic rubber production, the government had made it compulsory for the industry to buy first national rubber, and a tax (TORMB) was levied on imported rubber in order to equalise the prices of the national and imported product; the revenues from this tax were invested in the rubber-producing sector, but rarely in the 'wild' rubber sector. In 1990, however, in the context of the new liberalisation policies of the Brazilian government, and at the demand of the industrial sector, which wished to buy rubber at a lower price from abroad, the rubber tax was withdrawn.

Rubber tappers demonstrated in Brasilia asking for governmental support for rubber, and their demands were partially met in 1997, when the government agreed to sustain the subsidies for rubber production by paying the equivalent of the tax that was previously charged to industry (FoE/GTA, 1997; Allegretti, 1994). In principle, tappers thus receive a minimum price for their rubber, which is set up by Ibama. But this price has failed to keep up with inflation, buyers sometimes refuse to pay it, and given the overall decrease of the price of rubber, in some areas traders and factories have completely stopped buying rubber (Irving and Millikan, 1997). International factors also affect the current rubber crisis; the main reason why the price of rubber is decreasing is that imported rubber is very low-priced – Southeast Asian rubber costs US$1.60 per kg, whereas domestic rubber costs US$2.60 per kg. Besides the fact that the productivity of plantations is higher than that of the natural forest, this price difference is a result of national and foreign policies, such as the overvalued real, government subsides in Asia, and the low salaries of Asian workers (Irving and Millikan, 1997).

The rubber crisis has had different impacts on the various reserves, depending on the local external conditions and on the characteristics of the common-pool resources. In the Chico Mendes Reserve, many rubber tappers have left, especially the younger ones, and those who remained have switched to collecting Brazil nuts, which is now the main source of income in the reserve– in 1991, nearly 50% of the reserve inhabitants' revenue came from rubber (CNS, 1992). In the Alto Juruá Reserve, the large majority of the tappers have remained in the reserve because there are no job opportunities in the cities or elsewhere (local external factor); the main strategy of the Alto Juruá reserve inhabitants has been to switch to agriculture (de Almeida and Menezes, 1994; Irving and Millikan). In the Rio Cajarí Reserve, people also switched to other products, namely brazil nuts, açaí, and palm heart, depending on what abounds in each sector of the reserve, but in the Rio Ouro Preto reserve, given the lack of alternative products, the main strategy of its inhabitants has been until recently to abandon the reserve (Wawzyniak, 1994).

Another external factor influencing the reserves' inhabitants and their capacity to conserve the forest is the development of markets for other forest products besides rubber. There are over six thousand plants in the Amazon with recognised pharmaceutical and cosmetic uses (CNPT/Ibama, 1998), and in the Chico Mendes Reserve alone there are 64 different species with anti-inflammatory, anti-sceptic and analgesic properties, as well as other medicinal values (Ibama, c2000). The development of these products as alternatives to rubber depends on the international context, the level of external help rubber tappers receive, and their capacity to develop common property arrangements that both ensure the conservation of the forest and facilitate the commercialisation of such products. At the international level an important requirement is demand for such products and the prices they obtain in international markets. Incipient demand already exists and there are a few Brazilian and international companies interested in buying such products (Ibama, c2000). The extractive reserve inhabitants must also consider their competitors, who may charge considerably lower prices. This occurs, for example, with regard to the commercialisation of Brazil nuts, which at the expense of the labour force and the environment, are produced more cheaply in Bolivia[4].

Tappers need external help to identify demand for certain products and potential buyers, as well as to market their non-timber forest products. And as middlemen trade only rubber, and besides if they bought other products they would most probably pay very low prices, tappers need to strengthen their existing cooperatives and develop new ones. CNS estimates that for alternative economic activities to become viable in the reserves a period of 10 years is necessary. If the commercialisation of alternative forest products becomes profitable, the tappers' incentives to sell their stands or their tree to loggers would diminish, independently of the robustness of their common property regimes.

Initiatives such as PRODEX, the first governmental credit line aimed at supporting extractivism (Hall, 1997b) can also enhance the capacity of the reserve dwellers to ensure the conservation of their resources. One of the main problems the tappers have is their dependency on the middlemen. To develop co-operatives they need to realise that they can better the price of commercialising their produce, and must then engage in a collective action initiative. Yet without sufficient funds to buy the animals and pay someone to take the produce back and forth, even if communities are small, with homogeneous members and highly co-operative they will be unable to overcome the struggle that is obtaining the initial capital. Hence, the development of a credit line – a policy that is not directly related with common property – can influence the conservation of a commons.

Conclusion

At the time extractive reserves were created, several factors made clear that they were the most appropriate institution for the tappers, but ten years later, why is this still the case? The main reason is that the concept of extractive reserves continues to correspond to the rubber tappers' institutions. The extractive reserve decree states that resources are owned in common – the rubber tappers' informal institutions continue to be common property institutions and the problems that extractive reserves have are unrelated to the common ownership of the resource. There is no sign of a 'tragedy of the commons' occurring in, for example, the Chico Mendes Reserve. The problems the reserves' inhabitants face are of a different nature. The tappers abandon the reserves mainly because they cannot make a living in them; this is because the price of rubber is low and it is difficult to find alternatives to rubber extraction, which meet the environmental requirements of the reserve. However, even if the extractive reserves' legislation did not require the conservation of the forest, turning to unsustainable activities such as agriculture would fail to solve their livelihood problems, for in the long-term such activities are economically (as well as environmentally) unsustainable.

With regard to the tappers' lack of interest in the management of their resources, this might become a problem in the future, but it is not one at present. A more pressing issue is the development of cooperatives on the rubber estates and the joint management of other facilities. For tappers to fully engage in these activities, however, it is not necessary to change their property rights system, but to increase their autonomy and provide them with logistical and material help to enable them to set up such institutions. To solve these problems it is necessary that the tappers' property rights institutions continue to be recognised in the law and to receive state protection and support, which is what the extractive reserve concept is about.

The best proof that extractive reserves continue to be advantageous for commoners is that the high demand there is for extractive reserves in Brazil: Ibama is currently considering 22 requests for the creation of extractive reserves. Although tappers inside the reserves face a number of problems, those who lack the protection of the reserves are in a considerably worse situation. Notwithstanding the overall suitability of extractive reserves, and the improvement they represent for those communities whose rights are unprotected, altering some aspects of the co-management system (which pertains to Ibama's support framework, and not the extractive reserve model as such) would go a long way to enhance the capacity of the tappers to manage their own resources. Shifting the emphasis of the PP-G7 RESEX sub-project from developing robust regimes to further developing the tappers' economic alternatives and increasing their access to social services would reduce their incentives to leave the forest or sell the trees.

The second purpose of this chapter was to assess the capacity of the theory on common property regimes to explain the use of common resources. By and large, all the factors highlighted in the literature with regard to both the development and robustness of common property institutions could be discerned in the case of the rubber tappers' institutions. Particularly important for the development of the tappers' regimes were the potential of the resource to be held in common, the extractivists' access to information, and their autonomy to decide how to manage the resources. With regard to the capacity of these institutions to ensure the conservation of the forest, the definition of boundaries for the common-pool and the users was also fundamental. What the case of the tappers also shows, however, is that all these factors are contingent on the wider context; not only the wider political regime, but also attitudes of society towards resources and property rights institutions, legal issues not directly related to common property, and the level and type of help commoners receive from outsiders. Finally, poverty is a factor that also needs to be considered in the analysis of commons. Contrary to what has often been argued, poverty does not lead to resource destruction – the tappers have conserved their resources despite their poverty, and most destruction in Amazonia has been undertaken by wealthy cattle owners and mining companies rather than by poor farmers. Yet, economic hardship influences the decisions of individuals and should therefore be incorporated to current frameworks for analysing the use of common pool resources, especially in developing countries.

Notes

[1] There is very little information on the tappers' lives at the time, and it is not possible to make definitive statements from the interviews.
[2] As mentioned in Chapter 1, communal property requires that both the resource system and the resource units are common; on the other hand, *common* property involves common rights to the resource system but private rights concerning the resource units.
[3] At the time of the fieldwork, the programme was only beginning to be implemented and it is thus not possible to evaluate here the impact of the PP-G7 in the ERCM. However, in light of the evidence collected concerning the characteristics of the reserve and of the last independent appraisal of RESEX in 1997 (Irving and Millikan, 1997), the adequacy of the sub-project is assessed.
[4] See Assies 1997 for a detailed comparison between Brazil nut production in the two countries.

7 Supporting Local Resource Management

Introduction

As in many regions of the developing world, local communities in Amazonia depend to a higher or lesser extent on common resources for their livelihoods. For a long time, the established wisdom assumed that these resources were at risk of depletion, especially in the face of population increases or enhanced access to a market economy. In other words, local communities were considered unable to conserve their own means of livelihood. In the last few years, the outlook on community resource management has reversed, and now it is widely believed that local communities know better how to sustainably use their own resources, and that it is the unwelcome interference of outsiders that often leads to the destruction of common resources. There is substantial evidence to support this view: many local resources, including those of extractivist communities in Amazonia were threatened because of external actors, such as cattle ranchers, loggers, plantation owners, and small peasants.

The relationship between local communities and the external context is, however, not limited to a dispute over resources, in which local communities try to conserve them and external actors misuse them. This study has explored the evolution and sustainability of the Amazon rubber tappers' institutions in light of developments taking place in the wider context – local, national and international – showing how the external setting influences local communities' institutions both during their formation and once resource users have gained the right to manage their commons. Focusing on the Extractive Reserve Chico Mendes, the analysis has been framed by two questions: what influenced the development of common property regimes among the tappers, and the creation of extractive reserves? What affects the capacity of the reserves to ensure the conservation of the common resources? To understand the creation and sustainability of the rubber tappers' institutions the theory of institutional choice was used. Based on existing frameworks that identify the features of robust common property institutions and the factors that determine when commoners develop such regimes, a new framework including also a variety of external factors was developed (Chapter 1). As the theory of institutional choice has until recently tended to neglect external factors (with a few exceptions, see for instance Edwards and Steins, 1998), works on environmental action, which bring together several theoretical strands, such as political ecology, were also used.

The first time the rubber tappers attracted attention outside Amazonia was in the 1980s, when the international media published accounts of their struggle against cattle ranchers who were clearing the rubber estates to create pasture; in the following years, the tappers formed alliances with international environmental organisations, and in 1990 the Brazilian government legally recognised the extractivists' common property rights to the rubber estates and set up the first extractive reserves. Contrary to popular belief, however, the arrival of the cattle ranchers was not the first external factor to affect the tappers' institutions. This study has shown that from their inception, the tappers' institutions were shaped by the external context where they are embedded as much as by the characteristics of the tappers and their forests. The collapse of the Amazon rubber trade, for example, was what made possible the development of common property on the rubber estates, since with the abandonment of the estates by their private owners, tappers were left to their own devices and could decide how to manage their resources. Whereas some developments in the external context had negative impacts on the tappers' institutions, such as the policies that attracted ranchers to the Amazon region, others were largely positive, namely the return of democracy to Brazil in 1985 and the drafting of a new constitution in which environmental protection was given prominence.

The establishment of extractive reserves and their current institutional characteristics resulted from the interaction of internal factors, pertaining to the forest and its users, and external factors, originating in the wider context. Some of the most important determinants for the formation of extractive reserves were the characteristics of the forests, namely their indivisibility, the history of the rubber tappers, who had always shared the forest, even at the time of the rubber barons, the regional context, characterised by violent landed property rights conflicts, and international trends in society including interest in the environment, perceptions of the Amazon rainforest, and new understandings of the capacity of local communities to manage their resources.

Currently, the sustainability of the forest in extractive reserves depends to a large extent on the capacity of these institutions to ensure the conservation of the forest. This partly depends on the characteristics of these institutions, such as whether they ensure that the boundaries of the resource are well defined and protected. In the long term, it is equally important that the inhabitants of extractive reserves, tappers and other commoners, develop mechanisms to ensure their own sustainable use of the forest: rules on what is allowed, what is forbidden, and which activities have to be undertaken in common; monitoring mechanisms to ensure that all inhabitants comply with the rules, and conflict resolution mechanisms to enhance acceptance of the rules. In the short-term, however, other matters are more important. At present, the Extractive Reserve Chico Mendes, for example, is not threatened by the misuse of the forest by its inhabitants, but by their abandoning the forest in search of better economic alternatives. Such a problem is grounded in external developments, which have led to a substantial drop in the price of natural rubber. To reduce the outflow of extractivists from the reserves is necessary to develop economic alternatives for the rubber tappers. Some progress has already been made in this direction. In the context of the PP-G7 research into alternative forest products has been conducted and Ibama has initiated contacts with international companies interested in non-timber forest products. More, however, needs to be done, both to enhance the tappers economic alternatives and to facilitate the strengthening of their institutions.

Helping the rubber tappers

Ideally, the state should leave commoners sufficient autonomy to manage their own resources, but also support them when necessary; in practice, however, this is extremely difficult to achieve. In what follows, I do not pretend to advise on how to balance help and autonomy. The analysis of the rubber tappers' property rights institutions and of the wider context where they are embedded has, nonetheless, identified some issues in which there is a mismatch between the support the tappers receive and the characteristics of their institutions. Grounded on these observations, four suggestions are made on how to improve the effectiveness of the external support that the tappers receive.

First, external aid can be more effective if instead of attempting to implement common property institutions, it is confined to facilitate the development of such institutions and provide incentives for commoners to jointly manage their resources. Initiatives that help to catalyse co-operation among the forest dwellers include organising informal meetings between tappers in the area where they live, and increasing their access to information by presenting it in terms that they easily understand. In the

Chico Mendes and Alto Juruá Reserves, information should be provided inside the reserve, through regular visits by external actors, who could, for example, provide details on financial resources available to the tappers and what tappers from other estates are doing. In the Rio Ouro Preto reserve, by contrast, such information could be available in the cities, since many tappers either live there or travel regularly to the cities during the winter. If the forest dwellers have forums of discussion and easy access to information, they may attempt to develop cooperation in matters that they consider particularly important, such as trading of rubber; once they do this, the theory on common property predicts that if the conservation of the forest becomes a threat they are likely to address this problem expediently.

Second, rubber tappers should not be hurried into developing robust common property regimes because such a strategy is counterproductive. The development of cooperation requires time – resource users need to perceive that there is a problem, realize that something needs to be done about it, identify cooperation as the best solution, and then finally design effective institutions to address the problem. If the problem is not pressing – and in comparison with other issues, a future 'tragedy of the commons' is not – even if resource users perceive it, they are likely to have other priorities. Deciding to act, is also a slow process, especially for a population like the rubber tappers who has little experience of getting together to solve common problems; their institutions represent a form of passive rather than active cooperation, in the sense that tappers use their resources in common, but rarely manage them in common. With regard to the struggle against the ranchers, this study has shown that only a minority of the tappers participated in it. Besides, cooperating in a situation of crisis in which the members of the group need to fight against outsiders is different from defining long-term solutions to harmonise the actions of the group members themselves.

CNPT officials, acting in the context of the PP-G7, have tried to encourage the tappers to develop institutions to manage their resources and new facilities. One of the reasons for their limited success is that they have left the tappers insufficient time to adapt the new proposals to their own institutions, and hence extractivists have only nominally accepted their suggestions. Rubber tappers have lived in isolation and have survived under difficult conditions for many decades, and hence are characteristically reticent about new ideas. Moreover, some of the initiatives that Ibama suggested were to address problems that the reserve inhabitants had not previously felt. Overall there are also too many simultaneous new initiatives: to manage and represent the reserve, to manage the *seringal* and the new common facilities, to produce new products, and to develop cooperatives. The reason for outsiders making concurrent proposals is that all the issues they address seem to have equal priority. This, however, is not always the case. The conservation of the forest in the Chico Mendes Reserve, for example, does without the immediate strengthening of the tappers' institutions. With regard to the commercialisation of their produce, the development of small cooperatives seems more effective than the creation of cooperatives for the entire reserve; this was tried in the Chico Mendes, Alto Juruá, and Rio Ouro Preto Reserves, and serious problems occurred in all three cooperatives. By contrast, small cooperatives in the ERCM, like Primavera on Icuriã Estate (Assis Brasil) seemed to be particularly successful. And there are initiatives to address the tappers' problems that would go a great deal in reducing incentives to deplete the rainforest, without requiring the immediate development of robust regimes, such as extending the support for the price of rubber for some more years, and improving access to health and education.

Third, each extractive reserve is formed by a conglomerate of interdependent areas, which in some respects are similar (in all of them tappers hold their resources in common, depend on the forest for their livelihood, and have a common history), but that also present important differences; such differences need to be fully considered when designing aid for a reserve. In the Chico Mendes Reserve, the robustness of common property arrangements varies between rubber estates; consequently, the potential threats to the common pool resources differ between estates, and communities require help in different matters. On some estates, forest dwellers have hardly any contact with each other and are heavily indebted to the middlemen; what these communities most need is encouragement to meet more often and set up a cooperative. On other estates, tappers have already set up incipient common property arrangements and sometimes even small cooperatives; hence they need help mainly to strengthen what they have already rather than to establish new institutions. Similar differences exist within other reserves. In the Alto Juruá Reserve, there are two distinct areas, one around Rio Tejo, where tappers in the past revolted against their own patrons, and now regularly engage in joint activities, and the other around Rio Juruá where until recently hardly anyone participated in communal endeavours. In Rio Cajarí there are three different areas, mainly distinguishable because in two of them all transport is fluvial and in the other is done via a highway – such differences are bound to have implications in terms of accessibility and organisation of communities, and hence in the type of help commoners require.

Finally, the potential for developing nested enterprises in the extractive reserves should be given some thought. Once a sufficient number of communities is aware of the need to manage the forest, the development of the Chico Mendes Reserve as well as of other large reserves into a robust regime could be achieved by encouraging the development of a system of nested enterprises based on the local arrangements of each community. Given the ecological indivisibility of the rainforest, common property regimes based on small forest areas are unsustainable in the long term, yet the reserve cannot be managed as a single unit because its inhabitants do not perceive it as such. Formally, however, the reserve is managed as if it was a single common area formed by various rubber estates, instead of taking as a starting point the existence of *various* common areas that happen to be interdependent. A problem with this approach is that by encouraging the forest dwellers to participate in an anonymous group with which they do not feel affinity, it alienates them from taking responsibility for their natural resources. An alternative approach would thus be to articulate relations between the different institutions, after strengthening the common property regimes of the various communities.

The interaction of local, national and international factors in the sustainability of the common resources

A central argument of this study has been that the development and robustness of extractive reserves depend not only on internal factors, but also on the external context, including the socio-political context of the users, the legal setting, international trends in society, and world demand for forest products. Some aspects related to the external context that arose from the examination of the extractive reserves may be also relevant for policy on local resource management institutions in general.

The role of the state in relation to common property regimes is more than just 'facilitative', 'controlling' or 'indifferent'. An 'indifferent' state, which leaves commoners to their own devices and does not interfere with their resources can be as damaging to commoners as a controller state that takes over the commons. Regardless of the characteristics of the users and of the robustness of their institution, state ignorance of commoners' rights coupled with external pressure on the resource can lead to the break down of the common property regime and the destruction of the common resource. With regard to a facilitative state, provided that it refrains from imposing a utilisation of the resource that conflicts with that of the commoners, it can support joint users of a resource in a wide range of different ways: state agencies can *inter alia* protect co-owners against non-owners, provide arenas for conflict resolution, or supply information about the resource and other issues. Lastly, it should be noted that the state can do as much harm by promoting private access to resources which are common, as by encouraging common access to what so far has been privately used – in either case it can jeopardise robust property rights arrangements.

Besides the overall approach of the state, commons can also be influenced by the balance of power between different state departments at any one point in time. The state is rarely a monolithic entity; it is composed of a variety of agencies, and as the case of the tappers shows each of them can have a different approach to local resource management. In 1985 one state department, Pro-Memória, supported the tappers, while most of the other government agencies ignored them, and some branches actively injured; the police, for example, imprisoned rubber tappers who peacefully demonstrated against the ranchers' occupation of their forests and gave them insufficient protection against the ranchers' gunmen. Comparing the attitudes of Incra and Ibama towards respectively extractivist settlement projects and extractive reserves also suggests that state agencies vary in their approach towards common property regimes: both Incra's and Ibama's formal objectives concerning the tappers' newly created institutions are largely similar, but the serious problems the extractivist settlement projects have had and the relative success of the extractive reserves indicate that the government agencies' stances on common resource management differ.

With regard to international factors, it is not only communities depending on resources that are globally important such as the Amazon rainforest that are influenced by events in the international context. The inclusion of the international dimension in the examination of the rubber tappers' institutions originated in the fact that the tappers' forests are part of a resource whose importance is recognised worldwide. The research conducted, however, suggested that besides the ecological value of the Amazon rainforest, worldwide demand for forest products, the international community's interest in environmental matters and traditional populations, the political leverage of the concept of sustainable development, and macro-economic polices also influenced the tappers' institutions. These factors also impact on commoners who depend upon resources whose importance is not as widely recognised, such as the Pygmy communities of Central Africa, currently under threat from the operations of international logging operations, and the Chipko people in India, who counted on the interest of the international community on 'traditional peoples' to obtain support for their property rights' struggles, as well as other forest communities in Central America and Southeast Asia.

The examination of the external context in relation to community institutions should preferably not be limited to the features of the national and international setting at a given point in time. The robustness of common property institutions depends on the interaction of local actors with the external context over extended periods and the interests of local and international actors may vary over time. For instance, one of the factors that can contribute to the development of robust regimes is the possibility of local groups making alliances with international actors concerned with environmental issues, as exemplified by the role of the MDB campaign on the tappers' struggle for exclusive property rights. The motivations of local commoners and international environmental NGOs, however, are different; whereas local populations are interested in natural resources *for meeting* their needs, global actors (mainly from industrialised countries) are concerned with environmental conservation *after meeting* their needs. As Redclift (1992) comments when specifying differences between environmental interests in the North and in the South, in industrialised countries environmental issues are often seen as production versus environment, while in the South, in particular among the rural poor, environmental conservation is a requirement for production. In both cases, the conservation of the resource will be related to their other priorities: the tappers' interest in the forest, for example, depends on their economic alternatives, and public opinion concerning Amazonia in industrialised countries is contingent upon the economic situation at home, so that if there is a recession they will be less likely to contribute financially for a programme to conserve the rainforest. The factors that determine the interests of local and global groups are thus different and, accordingly their interests and priorities may not always coincide as they did in the late 1980s. The conservation of a commons will thus depend on the local users' capacity to ensure the conservation of their resources in a changing external context.

Commoners who depend on tropical forests for their survival are also likely to be affected by the conflicting views on forests of governments in the South and some actors in the North. As shown in Chapter 2, the debate on forests in the international arena has often been characterised by the issue of sovereignty and the global economy. On the one hand, developing countries where tropical forests are located, argue that they have the sovereign right to take decisions about their own natural resources, and that the current international economic system is partly responsible for deforestation; on the other hand, industrialised countries argue that, given the global impact of deforestation, they should have a say in the management of forests, and by and large ignore the environmental implications of the structure of the global economy.

The existence of this debate may be a positive factor for commoners because national and international support for local initiatives does not jeopardise national sovereignty and can, to a certain extent, be dealt with without altering the international economic system. Yet such a conflict can also have negative implications on forest people. Without the cooperation of national government agencies, international actors' influence on local communities is limited. They can help them to gain the support of their government, but cannot help them manage their resources unless there is a legal and social structure that supports local communities over time. National states alone, on the other hand, cannot address all issues affecting commoners, as many of them are related to wider economic trends beyond their control, such as demand for forest products. In Brazil, illegal logging is mainly fuelled by internal demand for tropical wood, in other regions of the world, however, such as Central Africa, most demand comes from the North – the problem of illegal logging thus requires international cooperation.

Looking ahead

Having developed a modified version of Ostrom's framework of common property to include national and international factors in the analysis, this study has contributed to understand the problems faced by extractive reserves, and has shed light on how local institutions interact with the wider context. Such work provides an insight into the complexity of the property rights arrangements of extractive reserves and on their potential for conserving the resource system. A brief overview of other extractive reserves in Amazonia included in this work suggested that many of the issues raised in relation to the Chico Mendes Reserve would be worth investigating in relation to the other reserves: do the state, *lideranças*, and the PP-G7 curtail the extractivists' autonomy in other reserves? Is the reserves' co-management system adequate for all reserves? To what extent can extractivists living in other reserves develop robust regimes in the long term? A detailed analysis of the various extractive reserves using the theoretical framework developed here could help to enhance the effectiveness of the aid they currently receive, and ensure that extractive reserves fulfil the expectations their creation rightly raised among forest dwellers and policy-makers.

The objective of the extractive reserves, however, is as much the conservation of the tappers' forests as it is the improvement of their socio-economic conditions. The links between property rights institutions and sustainable development, as opposed to resource management only, require further research. There is, on the one hand, a substantial body of literature on how issues of property rights relate to environmental conservation, but in the main it only addresses development issues insofar as sustainable management is a pre-requisite for rural and forest communities in the developing world. That is, the focus is on the sustainability of natural resources, but the rural poor need to develop strategies that in addition to conserving the natural resource base also increase their overall income. On the other hand, research in the field of development tends to neglect the potential of new institutional economics in addressing the rural poor survival strategies. Using interdisciplinary frameworks it is important to explore the creation, development and sustainability of local institutions in relation to their impact on rural poverty. And then to identify the impact that development and environmental policies (including international aid) have on these institutions, and how this translates into changes in the rural poor livelihoods.

Ultimately, the conservation of the forest in extractive reserves depends to a substantial extent on the level and type of support tappers receive from external agencies. Support for the rubber tappers in the context of the Pilot Programme as well as of other projects is temporary, as the objective of these projects is for tappers to become self-sufficient. Their aim is thus that tappers contribute to the conservation of the Brazilian Amazon without the help of external actors. From this perspective, a new question arises: to what extent can it be expected that the reserves' inhabitants and other commoners in Amazonia permanently contribute to the conservation of a globally important resource, without the international community carrying any of the costs?

Bibliography

Acheson, J.M. (1987), 'The Lobster Fiefs Revisited: Economic and Ecological Effects of Territoriality in Maine Lobster Fishing', in McCay, B.J. and J.M. Acheson (eds.), *The Question of the Commons: The Culture and Ecology of Communal Resources*, The University of Arizona Press, Tucson.

Acheson, J.M. (1989), 'Where have all the Exploiters Gone? Co-management of the Maine Lobster Industry', in Berkes, F. (ed.), *Common Property Resource: Ecology and Community-based Sustainable Development*, Belhaven Press, London.

Adams, W.M. (1990), *Green Development, Environment and Sustainability in the Third World*, Routledge, London.

Allegretti, M. (1979), 'Os Seringueiros: estudo de caso de um seringal nativo do Acre', Dissertação de Mestrado apresentada ao curso de Pós-Graduação em Antropologia da Universidade de Brasilia, Brasilia.

Allegretti, M. (1989), 'Reservas extrativistas: uma proposta de desenvolvimento da floresta amazônica', *Pará Desenvolvimento*, No.25 jan/dez 1989.

Allegretti, M.A. (1994), 'Reservas Extrativistas: Parámetros para uma Politica de Desenvolvimento Sustentável na Amazonia', in R. Arnt (ed.), *O Destino da Floresta: Reservas Extrativistas e desenvolvimento sustentável na Amazônia*, Relume-Dumará, Rio de Janeiro.

Alves, A.I. (1995), 'Descrição da Reserva do Rio Ouro Preto', in Arnt, R. (ed.), *O Destino da Floresta: Reservas Extrativistas e desenvolvimento sustentável na Amazônia*, Relume-Dumará, Rio de Janeiro.

Americas Watch (1991), *Rural Violence in Brazil*, Human Rights Watch, New York.

Anderson, T. and Leal, D. (1991), *Free Market Environmentalism*, Pacific Research Institute for Public Policy, San Francisco.

Arnold, J.E.M. and Campbell, J.G. (1986), 'Collective Management of Hill Forests' in Nepal: The Community Forestry Development Project', in National Research Council (ed.), *Proceedings of the Conference on Common Property Resource Management*, 24-26 April 1985, National Academy Press, Washington DC.

Arnt, R.A. (1992), 'The inside out, the Outside in' Pros and Cons of Foreign influence on Brazilian Environmentalism', *Green Year Book 1992*.

Artz, N.E., Norton, B.E. and O'Rourke, J.T. (1986), 'Management of Common Grazing Lands: Tamahdite, Marroco', in National Research Council (ed.), *Proceedings of the Conference on Common Property Resource Management*, 24-26 April 1985, National Academy Press, Washington DC.

Assies, W. (1997), *Going Nuts for the Rainforest: Non-Timber Forest Products, Forest Conservation and Sustainability in Amazonia*, Thela Publishers, Amsterdam.

Axelrod, R.M. (1984), *The Evolution of Co-operation*, Basic Books, New York.

Baden, J. (1977), 'A Primer for the Management of Common Pool Resources', in Hardin, G. and J. Baden (ed.), *Managing the Commons*, W H Freeman and Company, San Francisco.

Bandyopadhyay, J. (1992), 'From Environmental Conflicts to Sustainable Mountain Transformation: Ecological Action in the Garhwal Himalaya', in Ghai, D. and J. M. Vivian (ed.), *Grassroots Environmental Action: People's Participation in Sustainable Development*, Routledge, London.

Barbier, E. (1991), 'Tropical Deforestation', in Pearce, D. (ed.), *Blueprint 2: Greening the World Economy*, Earthscan Publications Ltd., London in association with the London Environmental Economics Centre.

Barbosa, L.C. (1993), 'The World System and the Destruction of the Brazilian Amazon Rain Forest', *Review* XVI 2:215-240.

Basilio, S.T.C. (1992), 'Seringueiro de Xapuri na luta pela Terra e Defesa da Floresta: projecto seringueiro, cooperativismo e educação popular', *mimeo*, PUC, São Paulo.

Batmanian, G.J. (1994), 'The Pilot Program to Conserve the Brazilian Rainforests', *International Environmental Affairs* 6(1): 3-13.

Bauer, D. (1987), 'The Dynamics of Communal and Hereditary Land Tenure among the Tigray of Ethiopia', in McCay, B. J. and J. M. Acheson (eds.), *The Question of the Commons: The Culture and Ecology of Communal Resources*, The University of Arizona Press, Tucson.

Berkes, F. (1986), 'Marine Inshore Fishery Management in Turkey', in National Research Council (ed.), *Proceedings of the Conference on Common Property Resource Management*, 24-26 April 1985, National Academy Press, Washington DC.

Berkes, F. (1987), 'Common-Property Resource Management and Cree Indian Fisheries in Subarctic Canada', in McCay, B. J. and J. M. Acheson (eds.), *The Question of the Commons: the culture and ecology of communal resources*, The University of Arizona Press, Tucson.

Berkes, F. (ed.) (1989), *Common Property Resources: Ecology and Community-Based Sustainable Development*, Belhaven Press, London.

Berkes, F. and Farvar, M.T. (1989), 'Introduction and Overview', in Berkes, F. (ed.), *Common Property Resources: Ecology and Community-Based Sustainable Development*, Belhaven Press, London.

Blauert, J. and Guidi, M. (1992), 'Strategies for Autochthonous Development: two initiatives in rural Oaxaca, Mexico', in Ghai, D. and J. M. Vivian (eds.), *Grassroots Environmental Action: People's Participation in Sustainable Development*, Routledge, London.

Blomquist, W., Schlager, E., Tang, S.Y. and Ostrom, E. (1994), 'Chapter 14. Regularities from the Field and Possible Explanations', in Ostrom, E., R. Gardner and J. Walker, *Rules, Games and Common Pool Resources*, The University of Michigan Press, Ann Arbor.

Brack, D. and Grubb, M. (1996), *Climate Change, A summary of the Second Assessment Report of the IPCC*, The Royal Institute of International Affairs, Energy and Environmental Programme, International Economics Programme, Briefing Paper No. 32 July 1996.

Bramble, B.J. and Porter, G. (1992), 'Non-Governmental Organisations and the Making of US International Environmental Policy', in Hurrell, A. and B. Kingsbury (eds.), *The International Politics of the Environment*, Clarendon Press, Oxford.

Branford, S. and Glock, O. (1985), *The Last Frontier Fighting over Land*, Zed Books Ltd. London.

Brock, L. and Hessler, S. (1993), 'The Colonisation of Amazonia: Constellation of Interests and Conflict Potentials', in Bothe, M., Kurzidem, T. and C. Schmidt (eds.), *Amazonia and Siberia: Legal Aspects of the Preservation of the Environment and Development in the Last Open Spaces*, Graham and Trotman/Martinus Nihhoff, London.

Bromley, D. (1985), 'Resources and Economic Development: an Institutionalist Perspective', *Journal of Economic Issues*, 3 (September): 776-96.

Bromley, D.W. (1989), *Economic interests and institutions*, Basil Blackwell, Oxford.

Bromley, D.W. (1991), *Environment and Economy: Property Rights and Public Policy*, Basil Blackwell, Oxford.

Bromley, D.W. and Cernea, M.M. (1989), *The Management of Common Property Natural Resources: some conceptual and operational fallacies*, World Bank Discussion Paper 57, World Bank, Washington DC.

Browder, J.O. (1988), 'Public Policies and Deforestation in the Brazilian Amazon', in Repetto, R. and M. Gillis (ed.), *Public Policies and the Misuse of Forest Resources*, Cambridge University Press, Cambridge.

Browder, J.O. (1989), 'Development Alternatives for Tropical Rain Forests', in Leonard, H. J. (ed.), *Environment and the Poor: Development Challenges for a Common Agenda*, Transaction Books, New Brunswick.

Browder, J.O. (1992), 'The Limits of Extractivism: Tropical forest strategies beyond extractive reserves', *BioScience*, 42(3): 175- 182.

Brown, K. and Rosendo, S. (2000), 'The institutional architecture of extractive reserves in Rondônia, Brazil', *Geographical Journal* 166, p. 201-228.

Buck, S.J. (1988), 'Cultural Theory and Management of Common Property Resources', *Human Ecology*, 17(1): 101-116.

Buck, S.J. (1998), 'Contextual Factors in the Development of Wildlife Management Regimes in the United States', Paper presented at the 8th

International Conference of the International Association for the Study of Common Property, Crossing Boundaries, June 10-14 1998, Simon Fraser University, Vancouver.

Burger, J., Ostrom, E., Norgaard, R.B., Policansky, D., and Goldstein, B.D. (eds.) (2001) *Protecting the Commons: a Framework for Resource Management in the Americas*, Island Press, Washington.

Campbell, C.E. in collaboration with the Women's Group of Xapuri, Acre, Brasil (1997), 'On the Front Lines but Struggling for Voice: Women in the Rubber Tappers' Defence of the Amazon Forest', *The Ecologist* 27-2: 46-54.

Cardoso, F.H. and Muller, G. (1977), *Amazônia: Expansão do Capitalismo*, Editora Brasiliense, São Paulo.

Carrier, J.G. (1987), 'Marine Tenure and Conservation in Papua New Guinea: Problems in Interpretation', in McCay, B. J. and J. M. Acheson (eds.), *The Question of the Commons: The Culture and Ecology of Communal Resources*, The University of Arizona Press, Tucson.

Carruthers, I. and Stoner, R. (1981), 'Economic Aspects and Policy Issues in Groundwater Development', *World Bank staff working paper* n.496, World Bank, Washington DC.

Cavalcante, O.P. (1993), 'A Polêmica em Torno do Conceito de Reserva Extrativista enquanto Atividade Econômica Sustentável', Monografia apresentada à Coordenação do Curso de Economia da Universidade Federal do Acre como requisito para a obtenção do grau de Bacherel em Economia, Orientador Prof. Reginaldo F. F. de Castela, Rio Branco, Acre.

CDEA (Commission on Development and Environment for Amazonia), c[1992] *Amazonia without Myths*, Inter-American Development Bank.

Chadwick, B., Bahar, H. and Albrecht, S. (1984), *Social Science Research Methods*, Prentice-Hall, inc. Englewood Cliffes, New Jersey.

Chaves, M.P.S.R. (1990), 'De "Cativo" a "Liberto": o processo de constituição sócio-histórica do seringueiro autônomo na Amazônia', *mimeo*, Projecto de Pesquisa, Universidade Federal da Paraíba, Centro de Humanidades, Mestrado de Sociologia Rural, Campina Grande – Paraíba.

Chopra, K., Kadekodi, G.K. and Murty, M.N. (1990), *Participatory Development: People and Common Property Resources*, Sage Publications, London.

Ciriacy-Wantrup, S.V. and Bishop, R.C. (1975), '"Common Property" as a Concept in Natural Resource Policy', *Natural Resources Journal* 15:713-27.

CNPT/IBAMA (1998), *Extractive Reserves: Business opportunities without environmental destruction, tropical forests cosmetics from extractive reserves*, Brasilia.

CNPT/IBAMA (c2000), Programa Piloto para a Proteção das Florestas Tropicais do Brasil – PPG-7, Projeto Reservas Extrativistas, Relatório Final da 1ª Fase 1995 –1999.

CNS (1992), 'Relatório Sócio Econômico da Reserva Extrativista Chico Mendes', Julho 1992, Rio Branco, Acre.

CNS (n.d.), *Diretrizes para um Programa de Reservas Extrativistas na Amazonia*, CNS, Acre.

Colchester, M. and Lohman, L. (1990), 'The Tropical Forestry Action Plan: What Progress?', *The Ecologist/The World Rainforest Movement*, Stuminster Newton, UK and Penang, Malaysia.

Cordell, J.C. and McKean, M.A. (1986), 'Sea Tenure in Bahia, Brazil', in National Research Council (ed.), *Proceedings of the Conference on Common Property Resource Management*, 24-26 April 1985, National Academy Press, Washington DC.

Cota, R.G. (1984), *Carajás: A invasão Desarmada*, Editora Vozes, Petropólis.

Cruz, W.D. (1986), 'Overfishing and Conflict in a Traditional Fishery: San Miguel Bay, Philippines', in National Research Council (ed.), *Proceedings of the Conference on Common Property Resource Management*, 24-26 April 1985, National Academy Press, Washington DC.

Danks, C. (1991), 'Extractive Reserves: A "Development Model" for Rain Forests? Distinguishing between the Means and the Ends', *Forestry*, 215, May 21, 1991.

de Almeida, M.W.B. (1994), 'As Reservas Extrativistas e o Valor da Biodiversidade', in Arnt, R. (ed.), *O Destino da Floresta: Reservas Extrativistas e desenvolvimento sustentável na Amazônia*, Relume-Dumará, Rio de Janeiro.

de Almeida, M.W.B. and Menezes, M. A. (1994), 'Acre – Reserva Extrativista do Alto·Juruá', in Arnt R. (ed.), *O Destino da Floresta: Reservas Extrativistas e desenvolvimento sustentável na Amazôni,a* Relume-Dumará, Rio de Janeiro.

de Assis Costa F. (1990), *A Discussão Brasileira sobre a "Internacionalisação" da Amazônia*, Jahrestagung der Arbeitsgemeinschaft Deutsche Lateinamerika-Forschung (ADLAF), Globale Vergesellschaftung und lokale Kulturen, Arbeitsgruppe, Wirtschaftskrise und Entschuldungsmechanismen: Lokale Entwicklungen – globale Lehren, Berlin.

de Onis, J. (1992), *The Green Cathedral, Sustainable Development of Amazonia*, Oxford University Press, Oxford.

de Paula, E.A. (1991), 'Seringueiros e Sindicatos: Um povo da floresta em busca da liberdade', Tese submetida como requisito parcial para a obtenção do grau de mestre em Desenvolvimento Agricola. Universidade Federal do Rio de Janeiro', Instituto de Ciencias Humanas e Sociais, Departamento de Letras e Ciencias Sociais, Itaguai, Rio de Janeiro.

Dean, W. (1987), *Brazil and the Struggle for Rubber: A Study in Environmental History*, Cambridge University Press, Cambridge.

Demsetz, H. (1967), 'Towards a Theory of Property Rights', *American Economic Review*, 62:347-59.

Diegues, A.C. (1992), *The Social Dynamics of Deforestation in the Brazilian Amazon: An Overview*, United Nations Research Institute for Social Development, Geneva.

Duarte, E.G. (1986), 'Conflitos pela Terra no Acre: a resistência dos Seringueiros de Xapuri', *mimeo*, Campinas.

Durrenberger, E.P. and Pálsson, G. (1987), 'The Grass Roots and the State: Resource Management in Icelandic Fishing', in McCay, B.J. and J.M. Acheson (eds.), *The Question of the Commons: The Culture and Ecology of Communal Resources*, The University of Arizona Press, Tucson.

Easter, K.W. and Palanisami, K. (1986), 'Tank Irrigation in India: An Example of Common Property Resource Management', in National Research Council (ed.), *Proceedings of the Conference on Common Property Resource Management*, 24-26 April 1985, National Academy Press, Washington DC.

Edwards, V.M. and Steins, N.A. (1998), 'The Role of Contextual Factors in Common Pool Resource Analysis', Paper presented at the 8th International Conference of the International Association for the Study of Common Property, Crossing Boundaries, 10-14 June 1998, Simon Fraser University, Vancouver.

Egger, P. and Majeres, J. (1992), 'Local Resource Management and Development: Strategic Dimensions of People's Participation', in Ghai, D. and J.M. Vivian (eds.), *Grassroots Environmental Action: People's Participation in Sustainable Development*, Routledge, London.

Ehernfield, D.W. (1972), *Conserving Life on Earth*, Oxford University Press, Oxford.

ELI (Environmental Law Institute) (1994), *As Reservas Extrativistas do Brasil: Aspectos Fundamentais de sua Implantação*, Environment Law Institute, Washington DC.

FAO (1994), *Forest resources assessment 1990*, FAO, Rome.

FASE/IBASE (1993), *Anais do Seminário de Estudos sobre o Programa Piloto para a Amazônia, Belem 1 a 4 Fevereiro de 1993*, Rio de Janeiro.

Fatheuer, T.W. (1994), *Novos Caminhos para a Amazônia? O Programa Piloto do G-7: Amazônia no Contexto Internacional*, FASE/SACTES, Rio de Janeiro.

Fearnside, P. (1985), 'Environmental Change and Deforestation in the Brazilian Amazonia', in Hemming (ed.), *Change in the Amazon Basin Vol. I Man's Impact on Forests and Rivers*, Manchester University Press, Manchester.

Fearnside, P. (1986), *Human Carrying Capacity and the Brazilian Rainforest*, Columbia University Press, New York.

Fearnside, P. (1989a), 'Extractive Reserves in Brazilian Amazonia: An opportunity to maintain tropical rain forest under sustainable use', *BioScience*, 39-6: 387-393.

Fearnside, P. (1989b), 'Deforestation in Brazilian Amazonia', in *The Earth in Transition: Patterns and Processes of Biotic Impoverishment*, Cambridge University Press, New York.

Fearnside, P. (1990), 'Environmental Destruction in Brazilian Amazonia', in Goodman, D. and A. Hall (eds.), *The Future of Amazonia: Destruction or Sustainable Development*, MacMillan, London.

Fearnside, P. (1997), 'Greenhouse Gases from Deforestation in Brazilian Amazonia: Net Committed Emissions', *Climatic Change*, 35: 321-360.

Fearnside, P. (1998), 'Deforestation Impacts, Environmental Services and the International Community', paper presented at the *Amazonia 2000 Conference'*, Institute of Latin American Studies, University of London, London 24-26 June.

Feitosa, M.L. (1995), 'Descrição da Reserva Extrativista Chico Mendes', in Murrieta, J.R. and R.P. Rueda (eds.), *Reservas Extrativistas*, UICN/CCE/CNPT, Cambridge.

Field, B.C. (1990), 'The Economics of Common Property: A Review of Two Recent Books – Proceedings of the Conference on Common Property Resource Management and The Question of the Commons, The Culture and Ecology of Communal Resources', *Natural Resources Journal*, 30 (Winter 1990): 239-252.

FoE (Friends of the Earth) (1991), *Mind the Gap! The Draft Pilot Program for the Brazilian Amazon: Considerations and Recommendations*, Summit of the Seven Most Developed Countries London, 15-17 July 1991, Friends of the Earth.

FoE/GTA (Friends of the Earth/Grupo de Trabalho da Amazônia) (1994), *Políticas Públicas Coerentes para a Região Amazônica; A harmonização de Politicas Públicas com os Objetivos do Programa Piloto.* Versao preliminar, documento de discussão da serie Mind the Gap!

FoE/GTA (Friends of the Earth/Grupo de Trabalho da Amazônia) (1997), *Políticas Públicas para a Amazônia: Rumos, Tendências e Propostas* Documento Apresentado à Reunião dos Participantes do Programa Piloto para a Proteção das Florestas Tropicais do Brasil, Manaus, 27-30 Outubro de 1997 Versão Preliminar.

Friedmann, J. and Rangan, H. (1993), 'Introduction: in Defence of Livelihood', in Friedmann, J. and H. Rangan (eds.), *In Defense of Livelihood: Comparative Studies on Environmental Action*, Kumarian Press, West Hartford.

Furley, P.A. (1990), 'The Nature and Sustainability of Brazilian Amazon Soils', in Goodman, D. and A. Hall (eds.), *The future of Amazonia: destruction or sustainable development*, MacMillan, London.

Gadgyl, M. and Iyer, P. (1989), 'On the Diversification of Common-Property Resource use by Indian Society', in Berkes, F. (ed.), *Common Property Resources: Ecology and community-based sustainable development*, Belhaven Press, London.

GoB/BIRD/CUE (Governo do Brazil/Banco International para a Reconstrução e Desenvolvimento/Comissão da União Europeia) (1994), 'Projeto Unidades de

Conservação de Uso Direto, Subprojeto Reservas Extrativistas, Programa Piloto para Proteção das Florestas Tropicais do Brasil, Brasilia.

Goldemberg, J. and Durham, E.R. (1990), 'Amazônia and National Sovereignty', *International Environmental Affairs*, 2(1): 22-39.

Gomes, M.E.A.C. and Felippe, L.D. (1994), 'Tutela Jurídica sobre as Reservas Extrativistas', in Arnt, R. (ed.), *O Destino da Floresta: Reservas Extrativistas e Desenvolvimento Sustentável na Amazônia*, Relume-Dumará, Rio de Janeiro.

Goodland, R., Ledec, G. and Webb, M. (1989), 'Meeting Environmental Concerns Caused by Common-Property Mismanagement in Economic Development Projects', in Berkes, F. (ed.), *Common Property Resources: Ecology and Community-Based Sustainable Development*, Belhaven Press, London.

Goodland, R.J.A. (1980), 'Environmental Ranking of Development Projects in Brazil', *Environmental Conservation* 7(1): 9-25.

Goodman, D. and Hall, A. (1990), 'Introduction', in Goodman, D. and A. Hall (eds.), *The Future of Amazonia: Destruction or Sustainable Development*, MacMillan, London.

Grima, A.P.L. and Berkes, F. (1989), 'Natural Resources: Access, Rights-to-use and Management', in Berkes, F. (ed.), *Common Property Resource: Ecology and Community-Based Sustainable Development*, Belhaven Press, London.

Gross, A. (1990), 'Amazonia in the Nineties: Sustainable Development or another Decade of Destruction', *Third World Quarterly*, 12(3), July 1990.

Grubb, M., Koch, M., Munson, A., Sullivan, F. and Thomson, K. (1993), *The Earth Summit Agreements: A Guide and Assessment*, Earthscan, London.

Hagemann, H. (1994), *Not out of the Woods yet – the scope of the G7 initiative for a Pilot Programme for the Conservation of the Brazilian Rainforests*, Verlag fur Entwicklungspolitik Breitenbach, GmbH, Saarbrucken.

Hall, A. (1987), 'Agrarian Crisis in Brazilian Amazonia: the Grande Carajás Programme', *The Journal of Development Studies*, 23(4): 522-52.

Hall, A. (1989), *Developing Amazonia: Deforestation and Social Conflict in Brazil's Carajás Programme*, Manchester University Press, Manchester.

Hall, A. (1990), 'Land Tenure and Land Reform in Brazil', in Prosterman, R., Temple, M. and T. Hanstead (eds.), *Agrarian Reform and Grassroots Development: Ten Case Studies*, Lynee Rienner, Boulder, London.

Hall, A. (1993), *Making People Matter: Development and the Environment' in Brazilian Amazonia*, Occasional Papers No.4, Institute of Latin American Studies, University of London, London.

Hall, A. (1996), 'Did Chico Mendes die in vain? Brazilian rubber tappers in the 1990s', in Collinson, H. (ed.), *Green Guerrillas: Environmental Conflicts and initiatives in Latin America and the Caribbean*, Latin America Bureau, London.

Hall, A. (1997a), 'Peopling the Environment: A New Agenda for Research, Policy and Action in Brazilian Amazonia', *European Review of Latin American and Caribbean Studies*, 62 (June 1997): 61-83.

Hall, A. (1997b), *Sustaining Amazonia: Grassroots Action for Productive Conservation*, Manchester University Press, Manchester.

Hames, R. (1987), 'Game Conservation or Efficient Hunting', in McCay, B.K. and J.M. Acheson (eds.), *The Question of the Commons: the culture and ecology of communal resources*, The University of Arizona Press, Tucson.

Hardin, G. (1968), 'The Tragedy of the Commons', *Science*, 162:1243-8.

Hardin, G. (1991), 'The Tragedy of the Unmanaged Commons', in Andelson, R.V. (ed.), *Commons without Tragedy: Protecting the Environment from Overpopulation - a New Approach*, Shepheard-Walwyn, London.

Hecht, S. (1983), 'Cattle Ranching in the Eastern Amazon: Environmental and Social Implications', in Moran, E.F. (ed.), *The Dilemma of Amazonian Development*, Westview Press, Boulder Colorado.

Hecht, S. (1985), 'Environment, Development and Politics: Capital Accumulation and Livestock Sector in Eastern Amazonia', *World Development* 13(6): 663-684.

Hecht, S. and Cockburn, A. (1989), *The Fate of the Forest: Developers, Destroyers and Defenders of the Amazon*, Penguin, London.

Hecht, S. and Schwartzman, S. (1988), 'The Good, the Bad and the Ugly: Amazonian extraction, colonist agriculture, and livestock in comparative perspective, *mimeo*, Graduate School of Planning, University of California, Los Angeles, CA, Environmental Defense Fund, Washington, DC.

Herrera, R. (1985), 'Nutrient Cycling in Amazonian Forests', in Lovejoy, T.E. and G.T. Prance (eds.), *Key Environments Amazonia*, Pergamon Press, Oxford.

Hilton, R.M. (1992), 'Institutional Incentives for Resource Mobilisation: an analysis of irrigation systems' in Nepal, *Journal of Theoretical Politics*, 4(3): 283-308.

Hirshon, R. (ed.) (1984), *Women and Property – Women as Property*, Croom Helm, London.

Homma, A.K.O. (1989), 'Reservas Extrativistas: uma opção de desenvolvimento viável para a Amazônia?', *Pará Desenvolvimento*, No.25 jan/dez 1989.

Homma, A.K.O. and Anderson, A.B. (n.d.), 'Oportunidades, Limitações e Estrategias para o Desenvolvimento Baseado no Extrativismo Vegetal na Amazônia', *mimeo*, CNS.

Hooker, A. (1994), 'The International Law of Forests', *Natural Resources Journal*, 34:823-877.

Huberman, A.M. and Miles, M.B. (1994), 'Data Management and Analysis Methods', in Denzin, N.K. and Y.S. Lincoln (eds.), *Handbook of Qualitative Research*, Sage Publications, London.

Humphreys, D. (1996), *Forest Politics - The Evolution of International Co-operation*, Earthscan Publications, London.

Hurrell, A. (1991), 'The Politics of Amazonian Deforestation, *Latin American Studies* 23:197-215.

Hurrell, A. (1992), 'Brazil and the International Politics of Amazonian Deforestation', in Hurrell, A.and B. Kingsbury (eds.), *The International Politics of the Environment: Actors, interests, and institutions*, Clarendon Press, Oxford.

Ianni, O. (1979), *A Luta pela Terra: Historia Social da Terra e da Luta pela Terra numa Área da Amazônia*, Vozes, Petropolis.

IBAMA (1995), *Plano de Utilização da Reserva Extrativista Chico Mendes – Acre*, IBAMA, Brasilia.

Ibama website (2001) http://www2.ibama.gov.br/resex/amazonia.htm

IBAMA/CNPT (1994), 'Roteiro para Criação e Legalização das Reservas Extractivistas Portaria do IBAMA N. 51 DE 11.05.94', in Murrieta, J.R. and R.P. Rueda (eds.), *Reservas Extrativistas*, CNPT/UICN, Cambridge.

INPE (1992), *Deforestation in Brazilian Amazonia*, São José dos Campos.

Instituto de Assuntos Culturais (1993), 'Relatório sobre Fortalecimento Gerencial das RESEX para Banco Mundial e CNPT/IBAMA, Consultants K.P.T Krauss, D.F.de Carvalho', Nov. 1993, Brasilia.

Irving, M.A. and Millikan, B.H. (1997), *Programa Piloto para a Proteção das Florestas Tropicais do Brasil, Projecto Reservas Extrativistas, Avaliação de Meio-Termo, Relatório de Avaliação Independente* Outubro 1997 Apoio: PNUD- Projecto BRA/92/043.

Jacobs, M. (1994), 'The Limits to Neoclassicism: Towards an Institutional Environmental Economics', in Redclift M. and T. Benton (eds.), *Social Theory and the Global Environment*, Routledge, London and New York.

Jaenicke, H. and Flynn, P. (1992), *Sustainable Land Use Systems and Human Living Conditions in the Amazon Region*, Commission of the European Communities.

Janzen, D.H. (1970), 'Herbivores and the number of tree species in tropical forests', *American Naturalist* 104:501-28.

Jodha, N.S. (1992), 'Common Property Resources: A Missing Dimension of Development Strategies, *World Bank Discussion Paper* 169, World Bank, Washington DC.

Johnson, O.E.G. (1972), 'Economic Analysis, the Legal Framework and Land Tenure Systems, *Journal of Law and Economics* 15:259-76.

Johnson, S.P. (ed.) (1993), *The Earth Summit, The United Nations Conference on Environment and Development (UNCED)*, Graham and Trotman/Martinus Nijhoff, London.

Jordan, C.F. (1985), 'Soils of the Amazon Rainforest', in Lovejoy, T.E. and G.T. Prance (eds.), *Key Environments in Amazonia*, Pergamon Press, Oxford.

Junk, W.J. and Furch, K. (1985), 'The physical and chemical properties of Amazonian waters and their relationships with the biota', in Lovejoy, T.E. and G.T. Prance (eds.), *Key Environments in Amazonia*, Pergamon Press, Oxford.

Keohane, R.O. and Ostrom, E. (eds.) (1995), *Local Commons and Global Interdependence, Heterogeneity and Co-operation in two Domains*, Sage Publications, London.

Kolk, A. (1996), *Forests in International Environmental Politics - International Organisations, NGOs and the Brazilian Amazon*, International Books, Utrecht.

Kovarick, M. (1995), *Amazônia-Carajás: Na trilha do saque*, Editora Anita Ltda., São Luís.

Kuehls, T. (1996), *Beyond Sovereign Territory, The Space of Ecopolitics*, University of Minnesota Press, London.

Kurien, J. (1992), 'Ruining the Commons and Responses from the Commoners: Coastal Overfishing and Fishworkers' Actions in Kerala State, India', in Ghai, D. and J.M. Vivian (eds.), *Grassroots Environmental Action: People's Participation in Sustainable Development*, Routledge, London.

Kwong, J. (1992), 'The Property Rights Alternative to Environmentalism, in Lewis, R. et al, *Rethinking the Environment*, Adam Smith Institute, London.

Lane, C. (ed.) (1998), *Custodians of the commons: pastoral land tenure in east and west Africa*, Earthscan Publications in association with the United Nations Research Institute for Social Development, London.

Langdon, S. (1984), 'The Perception of Equity: Social Management of Access in an Aleut Fishing Village', Paper presented at the Society for Applied Anthropology, Toronto, March 1984.

Lawry, S.W. (1990), 'Tenure Policy toward Common Property Natural Resources in Sub-Saharan Africa', *Natural Resources Journal* 30 (Spring 1990): 403-422.

Lélé, S.M. (1991), 'Sustainable Development: a critical review', *World Development* 19(6): 607-621.

Lewis, R. (1992), 'Clean Growth, The Free Market Way', in Lewis, R. et al, *Rethinking the Environment*, Adam Smith Institute, London.

Lipietz, A. (1995), 'Enclosing the Global Commons: Global Environmental Negotiations in a North-South Conflictual Approach', in Bhaskar, V. and A. Glyn (eds.), *The North The South and the Environment: Ecological Constraints and the Global Economy*, Earthscan, London and United Nations University Press, Tokyo.

Lisansky, J. (1990), *Migrants to Amazonia: Spontaneous Colonisation in the Brazilian Frontier*, Westview Press, Boulder.

Lovejoy, T.E. (1985), 'Amazonia, People and Today', in Lovejoy T.E. and G.T. Prance (eds.), *Key Environments in Amazonia*, Pergamon Press, Oxford.

Lovejoy, T.E. and E. Salati (1983), 'Precipitating Change in Amazonia', in Moran, E.F. (ed.), *The Dilemma of Amazonian Development*, Westview Press, Boulder Colorado.

MacPherson, C.B. (1978), 'The Meaning of Property', in MacPherson, C.B. (ed.), *Property*, Basil Blackwell, Oxford.

Mahar, D.J. (1989), *Government Policies and Deforestation in Brazil's Amazon Region*, World Bank, Washington.

Martin, K.O. (1979), 'Play by the Rules or Don't Play at All: Space Division and Resource Allocation in a Rural Newfoundland Fishing Community', in Andersen, R. and C. Wadel (eds.), *North Atlantic Maritime Cultures: Anthropological Essays on Changing Adaptations*, Mouton, The Hague.

Martine, G. (1990), 'Rondônia and the Fate of Small Producers', in Goodman, D. and A. Hall (eds.), *The Future of Amazonia: Destruction or Sustainable Development*, MacMillan, London.

Martins, J.S. (1980), 'Fighting for the Land: Indians and Posseiros in Legal Amazonia', in *Land, People and Planning in Contemporary Amazonia*, Proceedings of the Conference on the Development of Amazonia in Seven Countries, Cambridge 23-26 September 1979, Centre of Latin American Studies, University of Cambridge, Cambridge.

Martins, J.S. (1991), *Expropriação e Violência: a Questão Política no Campo*, Editora Hucitec, São Paulo.

McCay, B.J. and Acheson, J.M. (eds.) (1987), *The Question of the Commons: The Culture and Ecology of Communal Resources*, The University of Arizona Press, Tucson.

McCleary, R.M. (1991), 'The International Community's Claim to Rights in Brazilian Amazonia', *Political Studies*, XXXIX: 697-707.

McDonald, M.D. (1993), 'Dams, Displacement, and Development: A Resistance Movement in Southern Brazil', in Friedmann, J. and H. Rangan (eds.), *In Defence of Livelihood: Comparative Studies on Environmental Action*, Kumarian Press, West Hartford, Connecticut.

McElwee, P. (1994), *Common Property and Commercialisation: Developing Appropriate Tools for Analysis*, Masters Thesis in Forestry and Land Use, Oxford Forestry Institute.

McKean, M. (1992), 'Success in the Commons - a Comparative Examination of' Institutions for Common Property Resource Management, *Journal of Theoretical Politics*, 4(3): 247-281.

McKean, M. (2000), 'Common Property: What Is It, What Is It Good for, and What Makes It Work?', in Gibson, C.C., McKean, M.A and E. Ostrom (eds.), *People and Forests: Communities, Institutions, and Governance*, The MIT Press, Cambridge Massachusetts.

Melone, M.A. (1993), 'The Struggle of the Seringueiros: Environmental Action in the Amazon', in Friedmann, J.and H. Rangan (eds.), *In Defence of Livelihood: comparative studies on environmental action*, Kumarian Press, Connecticut.

Mendes, C. (1989), *Fight for the Forest: Chico Mendes in his own words*, Latin American Bureau (ed.), London (Adapted from *O Testamento do Homem da Floresta* edited by Candido Grzybowski, FASE, São Paulo, 1989).

Mendes, C. (1992), 'Peasants Speak: Chico Mendes – The Defence for Life', *The Journal of Peasants Studies*, 20(1): 160-176.

Menezes, M.A. (1994), 'As Reservas Extrativistas como Alternativa ao Desmatamento na Amazônia', in Arnt, R. (ed.), *O Destino da Floresta: Reservas Extrativistas e Desenvolvimento Sustentável na Amazônia*, Relume-Dumará, Rio de Janeiro.

Messerchmidt, D.A. (1986), 'People and Resources in Nepal: Customary Resource Management Systems of the Upper Kali Gandaki', in National Research Council (ed.), *Proceedings of the Conference on Common Property Resource Management*, 24-26 April 1985, National Academy Press, Washington DC.

Mishe, P. (1992), 'National Sovereignty and International Law', in Bilderbeek, S. (ed.), *Biodiversity and International Law - The Effectiveness of International Environmental Law*, IOS Press, Amsterdam.

Miyamoto, S. (1989), 'Preservação com Soberania' *Nossa América*, Março/Abril 1989.

MMA (Ministério do Meio Ambiente, dos Recursos Hídricos e da Amazônia Legal) (1995), *Os Ecossistemas Brasileiros e os Principais Macrovetores de desenvolvimento, subsídios ao planejamento da gestão ambiental*, Ministério do Meio Ambiente, dos Recursos Hídricos e da Amazônia Legal, Secretaria de Coordenação dos Assuntos do Meio Ambiente, Programa Nacional do Meio Ambiente, Programa das Nações Unidas para o Desenvolvimento, Brasilia DF.

Moran, E.F. (1981), *Developing the Amazon*, Indiana University Press, Bloomington.

Moran, E.F. (1990), *A Ecologia Humana das Populações da Amazonia*, Vozes, Petrópolis.

Musgrave, R.A. and Musgrave, P.B. (1973), *Public Finance in Theory and Practice*, Mc Graw-Hill, London.

Myers, N. (1989), *Deforestation Rates in Tropical Forests and their Climatic Implications*, Friends of the Earth, London.

National Research Council (ed.) (1986), *Proceedings of the Conference on Common Property Resource Management*, 24-26 April 1985, National Academy Press, Washington DC.

Nuggent, S. (1993), *From "Green Hell" to "Green" Hell: Amazonia and the Sustainability Thesis*, Occasional Paper no. 57, Amazonian Paper n. 3, University of Glasgow, Glasgow.

Oakerson, R.J. (1986), 'A Model for the Analysis of Common Property Problems', in National Research Council (ed.), *Proceedings of the Conference on Common Property Resource Management*, 24-26 April 1985 National Academy Press, Washington DC.

Olson, M. (1971), *The Logic of Collective Action*, Harvard University Press, Cambridge.

Ophuls, W. (1973), 'Leviathan or Oblivion', in Daly, H.E. (ed.), *Toward a Steady State Economy*, Freeman, San Francisco.

Ostrom, E. (1986), 'Issues of Definition and Theory: Some Conclusions and Hypothesis', in National Research Council (ed.), *Proceedings of the Conference on Common Property Resource Management*, 24-26 April 1985, National Academy Press, Washington DC.

Ostrom, E. (1987), 'Institutional Arrangements for Resolving the Commons Dilemma: Some Contending Approaches', in McCay, B.J. and J.M. Acheson (eds.), *The Question of the Commons: The Culture and Ecology of Communal Resources*, The University of Arizona Press, Tucson.

Ostrom, E. (1990), *Governing the Commons: the Evolution of Institutions for Collective Action*, Cambridge University Press, Cambridge.

Ostrom, E. (1992), 'Community and the Endogenous Solution of Commons' Problems', *Journal of Theoretical Politics*, 4(3): 343-351.

Ostrom, E. (2001), 'Reformulating the Commons', in Burger, J., E. Ostrom, R.B. Norgaard, D. Policansky, and B.D. Goldstein (eds.), *Protecting the Commons: a Framework for Resource Management in the Americas*, Island Press, Washington.

Ostrom, E., Gardner, R. and Walker, J. (1994), *Rules, Games and Common-Pool Resources*, The University of Michigan Press, Ann Arbor.

Ostrom, V. and Ostrom, E. (1978), 'Public Goods and Public Choices', in Savas, E.S. (ed.), *Alternatives for Delivering Public Services*, Westview Press, Boulder Colorado.

Paterson, M. (1992), 'Chapter 5: Global Warming', in Thomas C. *The Environment in International Relations*, The Royal Institute of International Affairs, London.

Pearce, D., Markandya, A. and Barbier, E.B. (1989), *Blueprint for A Green Economy*, on behalf of the London Environmental Economic Centre, Earthscan, London.

Peixoto, C.C.T. (1996), 'Amazônia, Meio Ambiente e Política Externa 1985-1992', Dissertação Apresentada ao IPR da Universidade de Brasilia Dezembro 1996.

Pejovich, S. (1995), *Economic Analysis of Institutions and Systems*, Kluwer Academic Publishers, London.

Peters, C.M., Gentry, A.H. and Mendelsohn, R.O. (1989), 'Valuation of an Amazonian rainforest', *Nature*, 339:655-656, 29 June.

Peters, P.E. (1986), 'Concluding Statement', in National Research Council (ed.), *Proceedings of the Conference on Common Property Resource Management*, 24-26 April 1985, National Academy Press, Washington DC.

Pinkerton, E. (1987), 'Intercepting the State: Dramatic Processes in the Assertion of Local Comanagement Rights', in McCay, B.J. and J.M. Acheson (eds.), *The Question of the Commons: The Culture and Ecology of Communal Resources*, The University of Arizona Press, Tucson.

Pinkerton, E. (1992), 'Translating Legal Rights into Management Practice: Overcoming Barriers to the Exercise of Co-Management', *Human Organisation*, 51(4): 330-341.

Pinto, L.F. (1982), *Carajás: O Ataque ao Coração da Amazônia*, Editora Marco Zero, Rio de Janeiro.

Porras, I. (1993), 'The Rio Declaration: A New Basis for International Cooperation', in Sands, P. (ed.), *Greening International Law*, Earthscan, London.

Porro, R. (1995), 'Alternativas Socialmente Desejaveis para a Implementação de Programas de Desenvolvimento Rural na Região Amazônica, *mimeo*, thesis draft, LBJ School of Public Affairs, Sate and Local Policy Making in Brazil, Professors Robert Wilson and Lawrence Graham, PA 882A.

Princen, T. and Finger, M. (1994), *Environmental NGOs in World Politics: Linking the Local and the Global*, Routledge, London.

Rangan, H. (1993), 'Romancing the Environment: Popular Environmental Action in the Garhwal Himalayas', in Friedmann, J. and H. Rangan (eds.), *In Defence of Livelihood: Comparative Studies on Environmental Action*, Kumarian Press, West Hartford, Connecticut.

Redclift, M. (1992), 'Sustainable Development and Popular Participation: A Framework for Analysis', in Ghai, D. and J.M. Vivian (eds.), *Grassroots Environmental Action: People's Participation in Sustainable Development*, Routledge, London.

Reeve, A. (1986), *Property*, MacMillan, London.

Revkin, A. (1990), *The Burning Season - Thė Murder of Chico Mendes and the Fight for the Amazon Rain Forest*, Collins, London.

Rich, B. (1994), *Mortgaging the Earth - The World Bank, Environmental Impoverishment and the Crisis of Development*, Beacon Press, Boston.

Richards, M. (1997), 'Common Property Resource Institutions and Forest Management in Latin America', *Development and Change*, 28:95-117.

Rowlands, I.H. (1992), 'The International Politics of Environment and Development: the Post-UNCED Agenda, *Millenium: Journal of International Studies*, 21(2).

Rueda, R.P. (1995), 'Evolução Histórica do Extrativismo', in Murrieta, J.R. and R.P. Rueda (eds.), *Reservas Extrativistas*, CNPT/UICN, Cambridge.

Runge, C.F. (1986), 'Common Property and Collective Action in Economic Development', *World Development* 14(5): 623-635.

Salati, E. (1985), 'The Climatology and Hidrology of Amazonia', in Lovejoy, T.E. and G.T. Prance (eds.), *Key Environments: Amazonia*, Pergamon Press, Oxford.

Samuelson, P.A. (1954), 'The Pure Theory of Public Expenditure', *The Review of Economics and Statistics*, xxxvi(4): 387-89.

Santos, R. (1984), 'Law and Social Change: the problem of land in the Brazilian Amazon', in Schmink, M. and C.H. Wood (eds.), *Frontier Expansion in Amazonia*, University of Florida Press, Gainsville.

Sautchuk, J. (1980), *Projeto Jari: a invasão Americana*, Editora Brasil Debates, São Paulo.

Sawyer, D. (1984), 'Frontier Expansion and Retraction in Brazil', in Schmink, M. and C.H. Wood (eds.), *Frontier Expansion in Amazonia*, University of Florida Press, Gainsville.

Sawyer, D. (1990), 'The Future of Deforestation in Amazonia: a Socio-economic and Political Analysis', in *Alternatives to Deforestation: Steps Towards Sustainable Use of the Amazon Rain Forest*, Columbia University Press, New York.

Schmink, M. (n.d.), 'Building Institutions for Sustainable Development in Acre, Brasil, *mimeo*.

Schmink, M. and Wood, C.H. (1992), *Contested Frontiers in Amazonia*, Columbia University Press, New York.

Schwartzman, S. (1989), 'Extractive Reserves: the rubber tappers strategy for sustainable land use of the Amazon Rain Forest', in Browder, J. (ed.), *Fragile Lands in Latin America: the search for sustainable uses*, Westview Press, Boulder Colorado.

Schwartzman, S. (1990), 'Social Movements and Natural Resource Conservation in the Brazilian Amazon', in Friends of the Earth (ed.), *Rainforest Harvest: sustainable strategies for saving tropical forests?*, Friends of the Earth, London.

Schwartzman, S. (1992), *mimeo*, Environmental Defense Fund, Washington DC.

Schwartzman, S. (1994), 'Mercados para Produtos Extrativistas da Amazônia Brasileira', in Arnt, R. (ed.), *O Destino da Floresta: Reservas Extrativistas e desenvolvimento sustentável na Amazônia*, Relume-Dumará, Rio de Janeiro.

Shoumatoff, A. (1991), *Murder in the Rain Forest: The Chico Mendes Story*, Fourth State, London.

Singh, K. (1994), *Managing Common Pool Resources – Principles and Case Studies*, Oxford University Press, Delhi.

Singleton, S. and Taylor, M. (1992), 'Common Property, Collective Action and Community', *Journal of Theoretical Politics*, 4:309-24.

Sioli, H. (1985), 'The effects of deforestation in Amazonia', in Hemming (ed.), *Change in the Amazon basin, vol I: Man's impact on Forests and Rivers*, Manchester University Press, Manchester.

Smith, R.J. (1981), 'Resolving the Tragedy of the Commons by Creating Private Property Rights in Wildlife', *CATO Journal*, 1:439-68.

Stevenson, G. (1991), *Common Property Economics: A General Theory and Land Use Applications*, Cambridge University Press, Cambridge.

Stocks, A. (1987), 'Resource Management in an Amazon Varzea Lake Ecosystem: The Cocamilla Case', in McCay, B.J. and J.M. Acheson (eds.), *The Question of the Commons: The Culture and Ecology of Communal Resources*, The University of Arizona Press, Tucson.

Sullivan, F. (1993), 'Forest Principles', in Grubb, M., M. Koch, A. Munson, F. Sullivan and K. Thomson (eds.), *The Earth Summit Agreements: A Guide and Assessment*, Earthscan, London.

Taylor, M. (1992), 'The Economics and Politics of Property Rights and Common Pool Resources', *Natural Resources Journal*, 32: 633-48.

Taylor, P.J. and Buttel, F.H. (1992), 'How do we know we have Global Environmental Problems? Science and the Globalisation of the Environmental Discourse', *Geoforum*, 23-3:405-416.

Thomas, C. (1992), *The Environment in International Relations*, The Royal Institute of International Affairs, London.

Thomson, J.T., Feeny, D.H. and Oakerson, R.J. (1986), 'Institutional Dynamics: The Evolution and Dissolution of Common Property Resource Management', in National Research Council (ed.), *Proceedings of the Conference on Common Property Resource Management*, 24-26 April 1985, National Academy Press, Washington DC.

Torres, H. and Martine, G. (1991), *Amazonian Extractivism: Prospects and Pitfalls*, Documento de Trabalho No. 5, ISPN - Instituto Sociedade, População e Natureza, Brasilia.

Tullock, G. (1977), 'The Social Costs of Reducing Social Cost', in Hardin, G. and J. Baden (eds.), *Managing the Commons*, W H Freeman and Company, San Francisco.

Turner, R.K., Pearce, D. and Bateman, I. (1994), *Environmental Economics – Elementary Introduction*, Harvester – Wheatsheaf, London.

van Ginkel, R. (1998), 'Contextualizing Marine Resource Use: a Case from the Netherlands', paper presented at the 8th International Conference of the International Association for the Study of Common Property, Crossing Boundaries, 10-14 June 1998, Simon Fraser University,Vancouver.

Viola, E.J. (1993), *A Expansão do Ambientalismo Multissetorial e a Globalisação da Ordem Mundial, 1985-1992*, Documento de Trabalho no. 16, ISPN – Instituto Sociedade, População e Natureza, Brasília.

Vivian, J.M. (1992), 'Foundations for Sustainable Development: Participation, Empowerment and Local Resource Management', in Ghai, D. and J.M. Vivian (eds.), *Grassroots Environmental Action: People's Participation in Sustainable Development*, Routledge, London.

von Behr, M. (1995), 'Descrição das Reservas Extrativistas do Extremo Norte do Estado de Tocantins, da Mata Grande e do Ciriaco', in Murrieta, J.R. and R.P. Rueda (eds.), *Reservas Extrativistas*, CNPT/UICN, Cambridge.

Wade, R. (1986), *Village Republics: Economics Conditions for Collective Action in South India*, Cambridge University Press.

Wade, R. (1987), 'The Management of Common Property Resources: Collective Action as an Alternative to Privatisation or State Regulation', *Cambridge Journal of Economics*, 11(2): 95-106.

WB/CEC (1991), *Pilot Program to Conserve the Brazilian Rain Forest Progress Report of the October 1991 World Bank/CEC Technical Mission to Brazil*, Washington, DC. November 13, 1991.

WB/CEC/GoB (World Bank/ Commission of the European Communities/ Government of Brazil) (1991), *Pilot Programme to conserve the Brazilian rain forest. Establishment of a Rain Forest Trust*, Washington, 13 November.

WCED (World Commission on Environment and Development) (1988), *Our Common Future*, Oxford University Press, Oxford.

Weinstein, B. (1983), *The Amazon Rubber Boom 1850-1920*, Stanford University Press, Stanford, California.

Weiss, E. (1989), *In Fairness to Future Generations: International Law, Common Patrimony and Intergenerational Equity*, The United Nations University, Transnational Publishers, inc. Dobbs Ferry, New York.

Welch, W.P. (1983), 'The Political Feasibility of Full Ownership Rights: The Cases of Pollution and Fisheries', *Policy Sciences*, 16:165-80.

Westerman, O. (1997), 'Empowerment for sustainable livelihoods in the Amazon: A socio-environmental analysis of the impacts of the extractive settlement projects Maracá I, II, III on people's livelihood', Masters Thesis, International Development Studies, Roskilde University Centre, 31 July 1997.

World Bank (1991), *The Forest Sector. A World Bank Policy Paper*, World Bank, Washington, DC.

WRI/UNDP/UNEP (1990), *World Resources 1990-91. A Guide to the Global Environment* Oxford University Press, Oxford.

WRI/UNDP/UNEP (1994), *World Resources 1994-95: A report*, Oxford University Press, Oxford.

Yadav, G., Roy, S.B., Dey, S. and Sarkar, S. (1998), 'An Assessment of Joint Forest Management in Regeneration and Management of Degraded Forest in West Bengal', Paper presented at the 8th International Conference of the International Association for the Study of Common Property, Crossing Boundaries, 10-14 June 1998, Simon Fraser University, Vancouver.